BUSINESS SECURITY

Over 50 Ways to Protect Your Business!

T. A. Brown

Crary Publications
Las Vegas, Nevada

Business Security was compiled by T. A. Brown. Copyright © 2004 T. A. Brown. All rights reserved. The use of the copyrighted articles in this collection has been authorized in every instance by the authors and/or their representatives. No part of this book may be reproduced, stored in a retrieval system, or transmitted by any means, electronic, mechanical, photocopying, recording, or otherwise, except brief extracts for the purpose of review, without the permission of the publisher. For further information contact Crary Publications, P.O. Box 42422, Las Vegas, NV 89116-2422 or check the website at www.BusinessSecurity.org

Disclaimer

These articles are written for general information purposes only and are not intended as a substitute for legal or technical advice. The specific facts that apply to your situation may make the outcome different than it would be for someone else. You should consult an attorney familiar with your particular situation and the laws in your state if you have questions. Neither the authors, publisher, nor any others involved in the making of this book assumes any responsibility for the use or misuse of information and sources contained in this book and specifically disclaim any liability or loss that may be incurred as a consequence of the use and application, directly or indirectly, of any information presented, including content found on suggested internet links, nor liability or responsibility for inadvertent errors or omissions. Readers accept all information herein without warranty, express or implied. Any slights of people, places, or groups are unintentional.

Edited by Dr. Charles Patterson
Cover Photo by Susan J. Stickle of www.hoofprints.ws
Cover Design by T. A. Brown & Cathi Stevenson

Publisher's Cataloging-in-Publication Data

Business security : protect your business against threats internally & externally / [compiled by] T.A. Brown.
   p. cm.
   Includes index.
   ISBN: 0-9743438-3-8
   ISBN: 0-9743438-9-7 (soft)
1. Business--Security measures. 2. Industries--Security measures. 3. Business intelligence. I. Brown, T. A.

HV6691 .B87 2004
658.4'73—-dc21

LCCN: 2003095795

ATTENTION CORPORATIONS, UNIVERSITIES, COLLEGES, AND PROFESSIONAL ORGANIZATIONS: quantity discounts are available on bulk purchases of this book for educational, gift purposes, or as premiums. Special books or book excerpts can also be created to fit specific needs. For information see the Order page in the back of this book or contact:

Crary Publications
P.O. Box 42422
Las Vegas, NV 89116-42422

Dedicated first and foremost to my children:
Dan, Diane, Jack, Tim, and Beth.

To my grandchildren, great-grandchildren and cousins.
And to my sister Linda Lou and her family.
All of whom comprise a very special family unit.

Also to Mr. Keith, a long ago trusted friend
who taught me humanity by his example.

Knowledge is of two kinds.
We know a subject ourselves,
or we know where we can
find information on it.
Samuel Johnson (1709—1784)

Table Of Contents

Introduction

Chapter One — Security 3
   Personal Security Measures 4
   Parking Lot Security 20
   Office Building Security 23
   Computer Security & Recovery 30
   Security and Company Culture 46
   Developing Approaches to Business Security 49
   Business Espionage Prevention 62
   Taping Phone Conversations 67
   Business Trip Security 69
   Disaster Management PS11 78

Chapter Two — Employee Screening 83
   Employee Screening Through Handwriting Analysis 84
   Background Checks in Canada 88
   International Resources 91
   Immigration Status Determination 98
   Pre-Employee Profiling 100
   Active Duty Military Verifications 103
   International Driver's Licenses 105
   Background and Credit Investigations 106

Chapter Three — Workplace 109
   Preventing Workplace Violence: Management Considerations 110
   K-9 Use in the Workplace 120
   Avoiding Computer Lockups 123
   Domestic Violence: A Concern for Employers? 128
   Crisis Intervention 132
   Preventing Sexual Harassment Lawsuits 141
   Responding to a Robbery 157
   Workplace Hostage Situations 161

Chapter Four — Fraud and Theft 181
   Terrorist Links to Commercial Fraud 182
   Insurance Fraud 187
   Do Not Become a Victim of a Scam 194
   Identity Theft Prevention and Checklist 197
   Reducing Shrinkage 200
   Interview or Interrogate 203
   The Enemy Within 207
   Tracking the Global Criminal 213
   Developing an Anti-Fraud Program 224
   How Some Terrorist Groups Raise Dirty Money in the USA 232

Chapter Five — Collections 237
    Collecting on Civil Judgments 238
    Post Office Box Process Service 245
    Unmasking the Mystery of Hidden Assets 246
    Asset Checklist.. 252
    Following the Money Trail... 256
    Enforcing Legal Judgments 260
    Nabbing Assets Before They're Eaten Up 263

Chapter Six — Information Gathering.. 267
    Public Records in Depth.. 268
    Information Databases 279
    Miscellaneous Tips 283
    Locating Companies in China... 297
    Corporate Intelligence Collection Process 299
    Finding a Reputable Private Investigator 304
    The Right to Privacy 308
    Dumpster Diving, ie: Garbology... 314

Acknowledgments 316

Author Contact Information 317

Index 321

Introduction

The purpose of this book is to provide you, the business owner, with the knowledge you need to arm, protect, and enhance your business concerns in today's ever-changing, challenging world of fraud, theft, lawsuits, corporate espionage, and electronic mediums.

The information contained in this book is a compilation of selected business-related articles that you can use to establish security measures for your business as well as a means to conduct background checks on your employees or applicants. The methods discussed in the articles are legal and ethical and are written by seasoned, experienced men and women in the field who work with this type of information on a daily basis.

Business Security will enable you to monitor what your employees are doing on the Internet, help you with policy-making decisions so that you can avoid sexual harassment lawsuits, give you the ability to conduct in-house pre-employment screening as an aid in reducing losses which result from employee theft and fraud, teach you how to check out your competition or check the immigration status of employees, and many other invaluable informational tools to use on a daily basis to ensure your business's security from both within and outside of your company, whether yours is a sole proprietorship or a large corporation. Employee theft and fraud are rampant in our society today. Did you know that approximately 60% of temporary employees have criminal histories? As a business owner who utilizes temporary employees, you cannot assume that "temp" agencies screen their employees. Background checks are essential to protect your business. This book will lead the reader to resources that may not have otherwise been considered as resources.

The tips included are vast, varied, and serve multiple purposes including, for example, a simple do-it-yourself method on how to check your office for electronic "bugs." There are hands-on methods as well as online methods of obtaining the information that your business needs to keep up in today's fast paced business community. Because of the rapidity of changes to information resources (due to new legislation, technological advances, etc.) some of the data contained in this book may have changed since this books' publication, but to the best of my knowledge all the information contained herein was up-to-date and accurate at the time this book was submitted for publication. The reader should keep in mind that the information contained herein and the security practices on which such information is based is constantly subject to new or changing facts or interpretations, governmental regulations, court rulings, or other events. Readers should verify the applicability of the information contained herein to their own situation on a state-by-state level.

Chapter One

Security

Personal Security Measures
By David P. Roberts

This extremely basic article is aimed at both the international diplomatic community and senior commercial enterprises, as well as those charged with their protection, as a substantive thought-engendering or aide-memoir tool.

By the same token, this article is formulated to be read, taught, and applied by respective family members and trusted employees—quite often the weakest links open to exploitation and exposure to security compromise—at the executive level, many of whom will be entering a cultural environment that is otherwise alien to them.

Security, inasmuch as it relates to personal protection, is a "constant learning" process. One has to try and teach the Principal one is charged with protecting, to accept a discernible degree of responsibility for their own security awareness and actions and ultimately arrive at a place where both Principal and Bodyguard agree about what their individual and unique security applications should be.

A Professional Review For High Profile Individuals

INTRODUCTION:
In current times and to an ever-increasing degree, it is essential for persons at risk to be constantly aware, even if subconsciously, of "Personal Security Measures" both at home and on the road—be it abroad on international excursions, or simply while traveling to and from work.

A thorough understanding and appreciation of the potential threats and risks will prove vastly beneficial in minimizing danger and risk. Apart from the application and use of certain electronic technical devices, coupled with acquired skills and discipline; the successful operations of these principles rely, in the main, on the practice of sound common sense.

Identifying or evaluating a potential threat, that is discussed here is designed to assist in developing the awareness factor that will minimize the level of exposure and/or danger to oneself and one's family.

The items contained herein are also promoted as a means to provide the VIP with the best possible security advice for him and his family members.

As a matter of interest, this unclassified publication is based on classified information which is imparted to every diplomat or consular official when first arriving to take up a new diplomatic posting in the British Isles.

There are topics here that in other publications develop beyond the basic insight provided in this article. Additional material is available which expands on many of the points contained in this article.

Only by knowing and regularly practicing the recommendations that follow, can the VIP hope to be able to apply such disciplines in his daily life and for the benefit of his family.

PERSONAL SECURITY MEASURES:
The VIP can rarely avail himself of an opportunity to fully relax his guard because the moment he does problems arise. However, a VIP's "mindset" can be trained to adjust to the potential for risk around him.

Even when asleep in a secure environment, part of the body's senses can remain alert to danger—a door opening, the flash of a torch, a dog barking, a waft of smoke—without interrupting a good night's sleep.

The following list of directives and guidelines should become so well entrenched in the subconscious of the VIP that, even without the application of conscious determined or dedicated thought, the processes will continue to operate just below the surface of the mind.

The VIP must **ALWAYS ANTICIPATE AND PLAN**

The continuous thought process should relate to:

1. WHERE and WHEN am I most vulnerable to attack?
2. Do I adhere to pre-determined routines? If you adhere to DAILY or REGULAR routines it is easy for somebody to anticipate your behavior pattern, one that facilitates selection of you or your family members for isolation and attack in circumstances that are most favorable to the attacker. Spontaneous and regularly varying activity will defeat the ability of an attacker to "anticipate and plan" against you.

You Must Retain The Initiative At All Times

GENERAL GUIDELINES

1. OBSERVATION & VIGILANCE:

Take note and be ever vigilant of your surroundings. Ask yourself, is there a strange or unusual vehicle parked in your street, outside your office or persistently in your rearview mirror. Sensitivity to your environment will enhance your capacity to looking after yourself.

2. ACTION:

REMEMBER - ACTION IS ALWAYS QUICKER THAN REACTION

If the potential attacker identifies in you a state of alertness or condition of awareness, you will retain the advantage. Attackers almost always look for:

- Weaknesses
- Vulnerability
- An escape route

Two notable and perhaps the most common exceptions to this profile are:

> The Fanatic. An example would be the suicidal religious zealot who drives an explosives-laden vehicle into the VIP's vehicle. In this case the area being visited might indicate risk, as might the nature of the visit or the coincidental presence of other targets.

> The Mentally Ill. There are people with an irrational and unpredictable motive to commit acts of violence. Almost any public appearance runs the risk of having a violently mentally ill armed person in the crowd. However, the same person may stalk his victim until an isolated position allows an approach, as happened in the John Lennon killing.

A less common example of an exception to this profile is the Professional Assassin, who would probably strike from a distance (sniper rifle, remote controlled bomb, booby trap etc.)

3. FAMILY CONSIDERATIONS:

As a means of neutralizing your security awareness, the "Softer Targets" in the form of your wife and children may be targeted. It should always be borne in mind that the information in this publication is not only for personal benefit, it is for the benefit of your family, colleagues and friends.

4. THREATS:

If you are the victim of threats, by whatever means, alert the Law Enforcement authorities immediately and advise your family members to upgrade their own security awareness. ALWAYS LET SOMEONE RESPONSIBLE KNOW WHERE YOU ARE OR INTEND TO BE.

5. DOMESTIC ARRANGEMENTS:

During holiday periods and other times of absence from the residence, cancel all deliveries, notify your local Law Enforcement Office (sometimes not appropriate), appoint a key-holder, and arrange for regular property checks for the period in question.

There are many commercially available devices that will give the impression of occupancy at all times, such as timer operated light/radio switches.

6. RAISING THE ALARM:

In the event of an incident developing or, your noting some suspicious activity, raise the alarm—call the Police or others capable of reacting by using the emergency numbers or electronic aids. Also, ensure that every member of the family is aware of how to initiate such activity independently if possible.

FIXED DAILY ROUTINES—THE "NEVER/ALWAYS" RULE:
Your observable patterns of behavior increase your vulnerability to attack.

1. NEVER leave your home at the same time each day.
2. NEVER take the same route to your place of employment or on the return journey.
3. NEVER walk alone, especially at night.
4. NEVER appear in public places on a regular basis.
5. ALWAYS ensure that your family adopts a similar attitude towards security disciplines.
6. ALWAYS ensure that your young children are supervised in traveling to and from school.
7. ALWAYS apply the "NEED TO KNOW" principle in communicating information about your personal and professional life.

POSTAL DELIVERIES:

Treat all parcels and packet deliveries with caution. Even an ordinary letter can contain sufficient explosive or incendiary material to cause considerable injury and/or damage.

Ruth First was killed by less than a matchbox full of plastic explosives packed in a "Walkman" sent by the South African Police Security Branch to her husband Joe Slovo.

CHECK LIST: Caution Recommended:

1. CHECK for postmarks from areas that could harbor those intending you harm.
2. CHECK the script of the address for recognition and/or correctness of presentation.
3. CHECK for stains—grease spots are caused by leakage of some explosive materials.
4. CHECK for smell—some explosives have the aroma of almonds or marzipan.
5. CHECK for weight—is the weight out of proportion to the contents or size of the package.
6. CHECK for size—do the contents appear equally distributed. Is the package uneven or misshapen?
7. CHECK for the protruding ends of wires or holes caused by wires.
8. CHECK the contents by holding it up to a bright light to determine the materials inside.
9. CHECK the contents for strengthening materials such as cardboard or metal.
10. CHECK for irregularities in the seal of the envelope flap. The presence of a complete seal on an envelope flap is unusual as is excessive application of sealing tape or stapling.

ACTION: Upon receiving a suspicious item, DO NOT PANIC.
Remember that it has traveled via the Postal Services and therefore been subjected to rough handling (unless hand delivered). Therefore, it should remain inert until the contents are determined.

Remove the item from the house to a safe area outside.

1. Place the item on a flat surface.
2. If the environment is not conducive to removal of the item, place it in a spare room farthest away from the remaining occupants and not in an area where they have to pass to evacuate the premises.
3. Evacuate the area and call the police/security on the emergency system.

4. With regard to any such delivery, if there is the least cause to be suspicious, in addition to the above-recommended course of action, never do the following.

THE "NEVER" RULE:

1. NEVER immerse the packet in water.
2. NEVER distort the packet by manipulation.
3. NEVER try to open or otherwise examine the contents. Ideally, there should be only one trained person who examines mail before it is distributed to other recipients. In any event, the basic practices mentioned above should be made familiar to all responsible family and/or staff members.

REMEMBER that sending materials that are offensive, noxious, contain dangerous and/or explosive materials, are threatening or otherwise identified as "Crank" letters, expose the person sending such materials to due process of the law.

Therefore, a suspect package must be considered of inestimable value to the Law Enforcement agencies tasked with location and prosecuting the offender. All care must be taken to preserve the best possible evidential value of such items.

It is incumbent upon the recipient of such items to minimize handling and secure the content for examination by those authorities. In the event of unknown or unexpected suppliers making or seeking to make a delivery:

1. Do not allow them access to your environment.
2. Seriously and fastidiously study the delivery details.
3. If the source of the delivery cannot be substantiated, do not accept the package.
4. Do not allow such deliveries to be deposited on the doorstep or windowsill.
5. Cultivate a rapport with the postman, milkman, and newspaper carrier. If there is any change in their regularity, treat such with suspicion until you can satisfy yourself as to the reason for any such deviation from the norm.
6. In all cases seek proof of identity from all unknown callers.

TELECOMMUNICATIONS - SECURITY AWARENESS:

1. The breaching of any implied integrity of security in respect to the telephone—fixed, mobile, or facsimile—is a relatively easy, albeit illegal, task for the skilled practitioner. In all

instances care should be employed when communicating on matters of confidentiality by this means.

A telephone is an essential line of communication that, if rendered inoperable, reduces your ability to call for assistance or for the police.

If the system fails for any reason, the defect should be immediately reported to the authorities for repair and the source of the fault determined. To do this, request from the operator an immediate "line check."

The source of the problem can thus be determined without delay and thereafter dictate any follow-up security activity. It is important to maintain a high degree of vigilance during the repair period and establish the bona fides of the engineer before allowing access to check or repair any defect.

2. NEVER position a telephone in a location where you can be seen answering it from external sources.

3. Do not note or delineate your telephone number on your own telephone handsets and certainly do not position the telephone in the view of callers to the property. And, as is more usually the case nowadays, there should always be a number of extension units, particularly in the bedroom.

4. Prepare a list of all essential and emergency numbers likely to be needed and keep a separate set available for immediate reference by all extension units.

5. Once again, and it is worthy of repetition, be aware of the very real fact that telephone conversations should never be considered as a secure means of communication. Therefore, there is a need to be extremely diligent in applying the "NEED TO KNOW" principle, especially when discussing travel arrangements, periods of absence from the home, movements of family members and/or domestic personnel.

ANONYMOUS - ANNOYANCE CALLS:
The frustration factor created by such attacks on the privacy of the individual cannot be sufficiently overstated.

There are some basic rules that should assist to defeat whatever the caller is seeking to achieve:

1. Keep the caller engaged in conversation as long as possible, regardless of how distressing the content of the call. While so doing, make special note of the following:

    a. Accent of Caller.
    b. Sex
    c. Probable age.
    d. Location from which call is being made by relying on ambient noises, i.e. listen beyond the voice.
    e. Whether coin or card operated telephone is being used.
    f. Date, time, duration and general content of call.

REMEMBER: Such callers are committing an offense punishable by law. You should strive to acquire as much evidence as possible.

2. Notify the Law Enforcement agencies immediately thereafter, giving full details in accordance with the above directions. Your evidence is vital and may be called for in any subsequent legal proceedings.
3. If the calls persist, consider using an answer phone facility to avoid having to personally answer each call. By monitoring the incoming calls, you can discriminate between those you wish to respond to or ignore. As an alternative, responding with a shrill blast from a whistle may have the desired shock effect.
4. It is always worth considering the facility of changing telephone numbers and becoming an "unlisted" subscriber. In any event, with the advances being made in respect of telephone technology, the benefits of an I.D. Caller System should not be overlooked. This facility will enable the telephone exchange staff to trace the location and unit from which the call was made.
5. Additional facilities exist, particularly in South Africa, the United States, the United Kingdom, and Europe, where all calls to your number can be routed via the exchange staff who will contact you for permission to connect the call to your line, after having first determined the identity of the caller and the location from where the call is being made.

SECURITY APPLICATIONS WITH RESPECT TO CHILDREN:
Children of tender years are particularly vulnerable, both inside and outside the domestic environment. Their innocence belies their ability to discern between genuine or suspicious activity, so the following rules should be positively considered:

1. Monitor their usage of the telephone since their innocent indiscretion may severely compromise all the steps taken by you in the development of good security codes of practice.
2. During the hours of darkness in particular, children should not be allowed to answer the door to any caller without being supervised by an adult.
3. While the manner in which contact with strangers should be handled outside the domestic environment is important, by the same token, when in the relaxed atmosphere of their homes, children should not be encouraged to greet strangers calling at the door of the residence with any the less caution.
4. Children's rooms should not be readily accessible to access by direct external means, viz: flat roof, fire escapes, garden trellis work, or large trees.
5. At the appropriate age children should be introduced to the use of the Emergency Telephone system and instructed accordingly.
6. At all times, your children should be able and encouraged to confide in you and notify you of their activities and whereabouts. Listen carefully to what they have to say. An innocent remark about such activities could be interpreted as having serious implications to an adult.
7. Only those persons who are competent, capable, qualified, and trustworthy should undertake the supervision of children when not with their parents. If babysitters are engaged, they should be from an accredited agency that utilizes personnel who are receptive to applying your security provisions and criteria and who have been vetted by the agency concerned.
8. The older the child, the more responsive they may be to accepting and applying security disciplines for themselves. They should be made fully acquainted with all matters of personal security.
9. Children of school age should never travel unsupervised and, at the very least, they should travel in groups. They too should not be allowed or encouraged to adhere to set routines and should never leave the school environment other than at pre-determined times.

SECURITY PROVISIONS FOR THE HOME:
In most places there exists within the offices of the local Law Enforcement Agencies, a Police "Crime Prevention" Advisory Department, whose assistance should be sought in advance of the installation of any electronic security equipment designed to enhance the security of your residence or that of your clients. Insurance companies may provide competitive discounted rates

once such facilities have been installed, the criteria being that only those installation agencies approved for such purposes are utilized.

Neighborhood Watch Schemes exist in most residential environments these days, and as far as the 00's are concerned, such public awareness and direct involvement can only escalate since necessity dictates that such should be the case. The value of such organizations in terms of their fight against domestic style crimes cannot be overstated. Working closely with those organizations and law enforcement agencies will markedly improve security conditions.

The following non-definitive list of domestic security measures are recommended for positive consideration:

1. All external doors should be fitted with mortise locks.
2. External doors and retaining frames should be of solid construction.
3. Approved "Chain" Systems should be applied to all external doors, if appropriate.
4. Opening windows at basement and ground levels should be fitted with locks, as should all accessible windows situated on other floor levels.
5. Suitable electronic systems can supplement the security of windows, with advice on their application always sought in advance.
6. During the hours of darkness movement to and around the premises to be protected can be readily identified by means of movement and/or heat detecting switches. In any event, the area should always be well lit with the appropriate equipment attached to the property shedding light outwards and onto anyone approaching the building. All to often, mainly for aesthetic effect, lights are placed in gardens, illuminating the property. Since such a scenario benefits anyone who wishes to approach the property covertly, this should be avoided.
7. Allocate keys carefully and never leave them outside the premises under any circumstances. In the event of a reported loss, change the locks accordingly. Resist the temptation to have the address and name of the occupant attached to any bunch of keys.
8. Allocate only those keys that are necessary to facilitate access.
9. Control the growth of shrubbery along all access routes and perimeters of the property. Do not provide cover for a potential assailant to hide himself or herself in anticipation of a physical confrontation.

10. Anticipate electrical faults and failures. Have available independent lighting sources on the premises, i.e. torches, candles etc. since they are essential for the security for your family.
11. Insist on advance notice of attendance by visitors—friends and relatives alike.
12. Satisfy yourself regarding the credentials of visiting workmen and ancillary staff members. Satisfy yourself about their status BEFORE opening any door. NEVER leave them alone on the property.
13. Unexpectedly late callers to the property should not be allowed unqualified access under any circumstances. Again, satisfy your-self about their credentials in advance of opening any external door.
14. Make a habit of checking all the internal security provisions in force at the residence in advance of retiring for the night.
15. Retain all waste bins and other waste disposal units within a secure area and be sure to destroy all items of personal correspondence prior to disposing of them. Anyone seeking to acquire information about the occupant of a particular residence will, in general, find all the information they want—and more besides—in the contents of domestic waste.

QUICK CHECK SECURITY RULES:

1. NEVER open a door without satisfying yourself as to the identity of the caller and his purpose. Security viewing equipment is readily and commercially available. Invest, because it could be a lifesaver.
2. If you have any cause whatsoever to be suspicious—CALL THE POLICE.
3. After dark ILLUMINATE FULLY all the approaches to the front door.
4. NEVER identify your keys by labeling them and strictly control duplicating.
5. NEVER seek to leave your keys in a "safe" location outside the house.
6. NEVER let external sources examine the interior of the residence during the hours of darkness—DRAW THE CURTAINS.

TRAVEL

There are some relevant considerations that MUST be brought into play in all instances involving travel, both with respect to arranging for the movements of a Principal and with respect to domestic plans.

The following list forms the basic minimal considerations for all security specialists and Personal Protection officers alike:

1. Strictly apply the "NEED TO KNOW" principle with regards to all specifics relating to travel arrangements.
2. Retain the facility to alter any such arrangements at the shortest notice.
3. Utilize as many different modes of transport and route variation as possible.
4. Try to travel in company whenever possible.
5. Do not draw attention by the way of ostentatious dress style or demeanor—strive to be as INCONSPICUOUS as possible.
6. Select a trusted employee to coordinate all travel arrangements.
7. Ensure that times and locations are coordinated and confirmed by direct contact at each port of call.
8. Always be aware of the surroundings, noting any suspicious vehicles or persons that may be present. Cultivate observational psychology techniques.
9. To avoid being followed rapid alterations in speed and direction can thwart mobile surveillance of one's movements. Spontaneous turns or stopping, negotiating a roundabout or circle, and making two complete circuits before exiting is a sure way to determine if one is being followed.

ALWAYS BE AWARE OF ONE'S ENVIRONMENT.

10. Keep your distance, when on foot, from any structural locations that would accommodate concealment, such as darkened recessed doorways, alleyways, overgrown or dense shrubbery, etc. Select a position on the footpath that facilitates time to react to anything spontaneous.
11. Consider utilizing a different name when making advance accommodation bookings.
12. Select popular areas when traveling by rail or coach. NEVER select an empty carriage compartment.
13. Air Travel, in spite of the inherent problems associated with security considerations, is still safer than travel by other means of public transport. BE AWARE OF YOUR ENVIRONMENT and do not expose yourself to risk at public transport facilities, if at all possible.
14. If possible—KEEP LUGGAGE IN VIEW AT ALL TIMES—Any evidence of tampering should be reported to the authorities without delay and PRIOR to personal examination.
15. When using taxicab facilities, enter the vehicle and ensure that it is moving BEFORE relaying the intended destination to the driver. That way nobody else except the driver will

know the destination. In any event, be alert to any possible breach in the "SECURITY OF INFORMATION" at all times.
16. Make a note of the license/permit number of the taxi and a mental note of the description of the driver.
17. In chauffeured vehicles, do not be averse to regular changes of seat positions.
18. Utilize the "NEED TO KNOW" principle at all times with regards to vehicular travel arrangements.

RULES & GUIDELINES FOR SECURITY DRIVERS.
By a combination of the application of all the security principles already highlighted, those that are to follow—common sense, awareness, and developed expertise—the safety of drivers and their passengers can be enhanced and assured.

EVALUATE THE RISK - APPLY THE PRINCIPLE RULES:

1. NEVER adhere to routine. It is a killer. Alter your routes and timings.
2. Utilize as many different vehicles as are available to you at random opportunities.
3. Secure the doors, trunk and hood at all times when parked. Utilize commercially available alarm equipment to alert you to tampering and, as a rule of thumb, neutralize the electronics of the vehicle.
4. Windows should remain closed at all times and never be opened more than sufficient to accommodate adequate ventilation in the absence of climate control facilities.
5. After a period of unsupervised absence from the vehicle, check the interior for persons lying in wait under objects.
6. If possible, do not leave the vehicle unattended. Select manned and secure garage facilities wherever possible.
7. Fit a locking gas cap.
8. Vary your use of gas stations—DO NOT become a creature of habit.
9. Check your communications regularly, particularly if a mobile phone is in use.

REMEMBER "Security of Information." A mobile phone is arguably the easiest communication system to infiltrate and compromise.

10. BE ALERT TO YOUR ENVIRONMENT. Vehicles and their occupants can be identified by plate registration numbers. When suspicious, record the number and inform the Law Enforcement agencies.
11. When in moving traffic, keep your distance from the vehicle in front and drive steadily. Once stationary, you become a

target. Your vulnerability is reduced if you are mobile, even if moving slowly.
12. Be extremely cautious in the event of spontaneous obstruction necessitating stoppage or slowing down of progress, such as occurs at accident scenes, etc. If obliged to stop by anyone other than a Law Enforcement Officer, do so as at great a distance as possible from the site. NEVER get out of the vehicle, and should the scenario develop into an attack, be ready to accelerate away. At all times, be aware of your total environment and, if possible, detour from your route in advance of being obliged to stop.
13. Select your routes in advance. Familiarize yourself with alter-native options. Select major roads where possible.
14. Change punctured tires only in safe and populated areas.
15. When stopped at traffic lights, ensure that you are positioned with sufficient space that you can take alternative evasive action to prevent yourself from being blocked in.
16. Hitchhikers are a definite "NO" at any and all times.
17. Police Officers are now aware that they must produce their identity cards before they expect motorists to expose themselves to questioning or for any other reason. INSIST ON IT. It is your right, and always take the name and number of the Officer concerned.
18. Do not leave items of value in view in the vehicle. The tidier the interior of a vehicle, the easier it is to identify tampering or unlawful entry.
19. Your garaging facility should be treated to the same security measures as your residence. KEEP IT LOCKED AT ALL TIMES.
20. After a period of absence from the vehicle, conduct the following checks before entering:

   a. Check the exterior of the car, including under the car, for attached articles.
   b. Check the interior for signs of entry and/or disturbances within.
   c. Check the rear wheel arches and both the front and rear of each wheel where it comes into contact with the road surface.
   d. Examine the hood and trunk for evidence of tampering.
   e. Examine the floor area around the vehicle for any signs of discarded or cut wires.

If for any reason you have cause to be suspicious, STOP IMMEDIATELY. UNDER NO CIRCUMSTANCES TOUCH OR

ATTEMPT TO START OR MOVE THE CAR. KEEP OTHERS AWAY FROM THE VEHICLE AND ALERT THE POLICE.

Internal vehicle checks take the following format:

1. Open the Door.
2. Check all areas of the driver's compartment.
3. Check the remaining internal areas.
4. Pay particular attention to the steering assembly.
5. Examine all internal compartments.
6. Examine the trunk.
7. Check the "on-board" equipment - tools, first aid kit, etc. Open the hood with care and examine the engine for signs of tampering and/or insertion or attachment of alien equipment.

THE SPACE BETWEEN THE ENGINE BLOCK AND BULKHEAD IS THE MOST VULNERABLE. IF IN DOUBT OR SUSPICIOUS, ALERT THE POLICE.

COMMONLY USED DEVICES:

1. Pressure contact switches—under wheels, seat covers, cushions, floor mats, pedals, etc.
2. Trembler or movement switches—designed to operate on mobility operation of vehicle.
3. Tilt Switches—designed to operate on negotiating uneven surfaces up or down gradients.
4. Wire operated switches—attached to doors, trunk or hood lids, or movable items of machinery.
5. Time operated switches.
6. Radio controlled devices.
7. Heat operated devices, normally attached to the exhaust manifold.
8. PEG Switch—which operates on the application of opposite forces.

Any device could be utilized as a single means of detonation, or a combination of these devices may be used to supplement or counter the failure of a single device.

INCIDENTS "ON THE MOVE":

If confronted with a developing incident while mobile, consider the following available courses of action:

1. Use your car/mobile phone to alert the authorities.
2. Consider an immediate detour, which should have been prepared in advance.

3. Facilitate your escape from any given situation. GIVE YOURSELF SPACE TO THINK AND ACT.
4. ALWAYS THINK BEFORE YOU ACT. REALISTIC APPRAISAL NEUTRALIZES THE RISK POTENTIAL.
5. Apart from the obvious safety reasons, you should automatically be constantly aware of your environment while driving, both to the front and the rear. KEEP YOUR DISTANCE.
6. In the event of a direct threat, KEEP YOUR VEHICLE MOVING. Any notes or recordings of your observations should be hidden or otherwise secreted in the vehicle. Draw as much attention to your predicament by means of horn, lights, hazard lights, and any other means available.
7. Record as many details as possible for evidential purposes.
8. Locate the nearest Police Office and report the incident in person.

The foregoing notes are designed to promote greater security awareness in every individual who reads them. They are not designed to induce alarmist reaction, nor undue cause for concern, but by engendering the thought processes that heighten security awareness and by developing an individual's common sense about the application of these recommendations and guidelines, it is hoped that people will respond sensibly and logically, thereby enhancing their own security considerations within, and relative to, their own environment and professional terms of reference.

All due regard must be paid to everyone's status and station in life, relative to their individual exposure to risk and of course, that of their respective family members.

IN THE FINAL ANALYSIS, WE ARE ALL RESPONSIBLE FOR OUR OWN SECURITY

Written by the internationally famous bodyguard who was once the bodyguard for Whitney Houston and whom the movie "The Bodyguard" was based upon. Former: Supervisory Officer, Royalty & Diplomatic Protection Department. Metropolitan Police, New Scotland Yard, London.

Parking Lot Security
By Gary R. Cook

Security for parking lots seems to be getting a lot of press lately. If you look at the statistics, roughly 80% of the criminal acts at shopping centers, strip malls, and business offices occur in the parking lot. Lawyers make a good living off liability cases based on a lack of sufficient security measures or not taking "reasonable care" in the protection of employees and customers against criminal threats. The lawsuits often revolve around lack of sufficient lighting, surveillance, and response. Once crime takes a foothold in an area it is difficult to break the trend, but there are some things you can do that can improve security, deter crime, reduce potential liability, and make your customers feel safer. It's interesting to note that where parking lot security has been implemented, customer use has actually increased because they feel safer. Increased customer use means increased profit, which can be used to justify the increased cost, related to any security improvements.

Lighting
Security lighting is used to increase effectiveness of guard forces and closed circuit television by increasing the visual range of the guards or CCTV during periods of darkness or by increased illumination of an area where natural light does not reach or is insufficient. Lighting also has value as a deterrent to potential individuals looking for an opportunity to commit crime. Normally security lighting requires less intensity than working areas. The exception is at normal doorways. Exterior lighting for areas such as parking lots is required to ensure a minimum level of visibility when guards are used to perform inspection duties of the protected area. Guards and CCTV surveillance systems must be able to identify badges, people, and guards at gates, observe activity, inspect vehicles, observe illegal entry attempts, detect intruders in the protected area, and observe unusual or suspicious circumstances. Each parking lot presents its own particular problems based on physical layout, terrain, atmospheric conditions, and security requirements.

The goal of direct illumination is to provide a specified intensity throughout the area for support of guard forces or CCTV, provide good visibility for customers or employees, and have a minimum of glare. The most severe problem is illuminating the small narrow "corridors" formed by adjacent parked cars. In order to get some light into these areas, it is recommended that any point in the entire parking lot be provided with illumination from at least two and preferably four lighting (pole) locations and with the lights mounted at a minimum height of 20 ft. (lowest value on the pavement should not be less than one fourth of the recommended average). The minimum recommended illumination level for the barest sight essentials on the parking lot proper is one-foot candle (average maintained horizontal to the surface) for self-parking areas and two-foot candles for attendant parking areas (because of liability and potential damage to automobiles). Where additional lighting for business attraction or customer convenience is a consideration, five-foot candles and higher is often used.

Illumination levels at entrances, exits, loading zones, and collector lanes of parking areas should not be less than twice the illumination of the adjacent parking area or the adjoining street, whichever is greater. Lighting poles should be mounted along the parking barriers and outside boundaries. Lighting requirements for CCTV are considerably lower than those required for direct visual observation depending on the type of system selected. CCTV cameras must be oriented so the rising or setting sun, automobile headlights, or reflections from parking lot luminaries do not blind them.

Layout
The layout of a parking lot can sometimes provide an advantage for natural surveillance, CCTV coverage, and structured surveillance and response. Parking lots for retail centers are unique because there is no way to control who has access as opposed to business parking areas that can and do control access if the potential risk to employees justify it. For layout of parking for retail centers, first look at the potential parking sequence and determine if there is a way to increase natural surveillance. Normally retail patrons who arrive early also leave early, leaving late arrivals the less secure (farther away from the store front and traffic flow) parking spaces. Since these late arrivals also are usually the last to leave, they are also the most vulnerable to crime. By rerouting incoming and outgoing traffic through the parking lot to pass by the more remote areas, natural surveillance is increased and criminal opportunity is reduced.

Positioning of the layout to increase the effectiveness of CCTV surveillance at the parking areas can also be cost effective. Parking perpendicular to the line of sight and CCTV coverage reduces the criminal value of hiding between cars waiting for potential victims. Walking corridors between cars at strategic locations also concentrates foot traffic and increases natural surveillance by retail patrons.

Response

Surveillance without potential response provides little increase in system trust by customers. It is not uncommon in high crime or remote areas to install Emergency Call Stations that can be used to call security forces or police to an emergency situation. Availability of these call stations for use by customers observing a crime in progress or by victims who are threatened provides a considerable increase in comfort level for employees and customers. These systems provide immediate voice contact (with security forces), alarm (to attract attention), and light signal (quick location of trouble spot).

The use of radio equipped bicycle patrols and golf carts are also gaining popularity to provide quicker security response and service for customers (ambulance, AAA, locksmith, etc.) as well as a more consistent and frequent presence. A visual reminder of territoriality is sometimes all that is necessary to deter crime. The more consistent the visual reminder, the higher the deterrent level. This method was a factor in the success of security at the 1984 Olympic Games in Los Angeles. Every attendant, volunteer, and security person was dressed in a bright easily recognized uniform. They were a constant and overwhelming reminder that security was important and that someone was always near to sound the alarm and summon help.

Risk

Implementation of these measures is a function of risk and risk management. I strongly recommend that for this and any other security improvement plan, a risk analysis be performed that will help convince management that security improvements are justifiable and cost effective. The asset at risk in parking lot security is the personal property and well being of your customers and employees. How much you are willing to spend to protect them and keep them as customers is a function of their value to you as an organization. If there are other shopping centers to shop at or other businesses to work for and your customers and employees will go there if they feel safer, the cost is apparent.

Gary R. Cook, P.E., is a registered professional engineer in the State of California, the owner of *Security Design Sciences* in Ventura, CA, and the publisher of *Security Design Newsletter*, a free quarterly publication focusing primarily on physical security issues.

Office Building Security
By Ralph Witherspoon

There are more than a million office buildings in the United States, and more and more Americans are spending a significant part of their lives working in them. Corporations and businesses increasingly house their most important assets—their employees and their printed and electronic information—in office buildings.

And just as bank robbers rob banks because that is where the money is, many criminals go to office buildings today to steal, rob, rape, and spy because that is where their potential victims are located.

This brief article will provide the reader with guidance on some of the basic issues concerning securing office buildings in America today.

First, note that nothing discussed here will prevent the type of destructive attack we saw on September 11th. Only the government can do that. What is addressed here is ground based attacks, both terrorist and criminal.

Both government and private commercial buildings may be targeted. Within the U.S. alone, the World Trade Center (WTC) buildings were previously targeted in 1993 with a truck bomb in the underground garage; a subsequent plot to bomb a New York City bridge and underwater tunnel was thwarted; the Murrah Federal building in Oklahoma City was destroyed in 1995 by domestic terrorists with a truck bomb; and letter bombs and anthrax letters have been sent to both government and corporate offices. The traditional threats and building security issues remain, but new (covert) terrorist threats have assumed added importance.

These terrorist threats originate not only in the international arena, but also come from a constantly changing cast of environmental, animal rights, anti-abortion, neo-nazi, anti-big business, anti-globalization, and other similar groups and activists. Some individuals in these groups have demonstrated that they are not averse to using violence to further their cause.

One Size Doesn't Fit All
Single-tenant buildings (occupied by a single tenant involved in one type of business) offer different security issues and risks than does a multi-tenant / single-use building, such as a medical office

building with many tenants all in the same type of business. A still different situation exists for a multi-tenant / multi-use building which may have many different tenants doing many different things, including offices, retail operations, public utility or agency offices, operating parking facilities, and more.

How Likely a Target Are You?
There is no "cookie-cutter" plan that will secure each and every type of office building, high-rise or low-rise. Management of each will have to identify its own security needs, starting by conducting a risk assessment. How likely a target you are perceived to be by criminals and terrorists will, in large part, determine how likely you are to be attacked. And how you are perceived depends, in part, on what visible security measures you have in place and how effective they are perceived to be.

In a commercial office building, security risks may include murder, robbery, rape, assault, theft, commercial espionage, arson, vandalism, bomb threats, and sabotage, to name but a few. The heavy concentration of people and property, coupled increasingly with "open" floor plans, make modern high-rise buildings susceptible to these type threats. Plus, the always-present life-safety risks include fire, explosion, and natural disasters. And today the possibility of a *covert* terrorist attack has to be considered by many facilities (an armed attack with multiple attackers would require a police or military response).

Management responsible for securing any commercial office building should first assess the risks to the building and its tenants. A survey should be made of all tenants to ascertain what type of business each is conducting, what significant business assets are present, and which, if any, may constitute an increased risk to the building and other tenants from criminals, political activists, political terrorists, etc.

Additionally, talking with tenants also helps get them involved in the process so that the building's security plan is not designed in a vacuum. The tenants and their employees will have to live with the resulting security plan, and if they won't cooperate with its measures, it won't work. The assessment should include a review of any known past crimes in the building along with an evaluation of crime in nearby office buildings and in the immediate area. Local law enforcement will usually provide data on these situations.

How likely a crime target any specific building is depends on the perceived gain to the perpetrator, balanced against his perceived risk of apprehension or defeat. For a criminal it is usually monetary gain to be realized. For the industrial or commercial spy it is the corporate secrets and other sensitive or confidential information. For the sexual predator it is the women who work in the building or use the garage. For a terrorist it may be the media attention gained by the destruction of lives and property. This is especially true if the building is well known, such as the WTC, and/or is perceived to be a symbol of America, such as the Bank of America, or is the home of a well-known, prosperous, or controversial corporation or group such as the World Trade Organization (WTO) or the World Bank (WB).

In almost all cases the criminal will have to first access the building itself to reach his target(s). International or domestic terrorists using a bomb may only have to get "close," however. Usually the criminal or terrorist prefers a "soft" or easy target if it will achieve his objective.

Based on an assessment of the likelihood of certain security events happening, e.g. theft, assault, robbery, bombing (or damage from bomb blasts at nearby "high risk" buildings, such as government buildings, etc.), a level of risk can be determined by management.

Security Survey
Next, a security survey (an exhaustive physical examination of the building, including a review of its security processes, policies, and procedures) should be conducted. Local laws and codes pertaining to security measures, fire codes, and building evacuation requirements should also be reviewed. Based on the identified risks, plus any identified gaps or shortcomings in security (vulnerabilities), management can start to develop an overall security plan and to identify cost-effective counter-measures to provide maximum deterrence to criminals and terrorists alike.

Where management does not have qualified expertise on staff, or such expert staff is not readily available to conduct a survey due to other commitments, an independent non-product affiliated security consultant should be retained.

Parking and Adjacent Spaces
In most downtown office building locations management probably cannot control or prevent vehicles from stopping or parking on the public street next to their building. They may or may not be able to control the alleyways next to it. As a result, if the threat of a bombing exists, the risks from a car or truck bomb increases. Decorative concrete barriers or bollards can be used to provide some separation between vehicles and the building, thereby reducing the blast effect. While the risk of bombs to most non-government or other "high-risk" office buildings is usually not high,

it is not non-existent. Individual building risk may also be increased by the presence of foreign government consulates, highly visible or controversial tenants, or individual federal or state government offices.

Special attention should be given to any underground, adjacent, or attached parking spaces or garages. These are not only frequent targets of criminals committing theft, robbery, rape, and carjacking (and a source of many lawsuits against building owners and managers), but they also have been used in some cases to place a vehicle bomb next to the building, or in a garage under the building (as in the 1993 WTC bombing).

Vehicles parking next to the building should be restricted and controlled, if at all possible. If the building or its tenants are "high risk," such parking should be prohibited, or moved at least 100 feet away from the building. The destruction of the Murrah Federal Building in Oklahoma City illustrates the damage that can result from this type car or truck bomb.

If underground parking is permitted, it should only be granted to known tenants, and, depending on risk, it may be necessary that such vehicles be inspected or searched upon entry. Access of trucks to underground loading docks should be strictly controlled, with document and vehicle inspections of all trucks made _prior_ to their entry.

Interior garage lighting should be a _minimum_ of five foot-candles (55 lumens) throughout the garage, 24-hours per day. Sunlight seldom enters and cannot be relied upon. Inspection points require at least 15-20 foot-candles (165B220 lumens) of illumination. Interior walls should be painted with a glossy white paint to increase light reflection off the walls. Pillars should be painted in contrasting colors.

At a stand-alone office building with adequate "green space" surrounding it (usually in the suburbs or smaller towns) and with low traffic levels, vehicles can be inspected and cleared at the entryway to the property. Speed bumps and road curves can be utilized to slow vehicle traffic and to direct it away from the main building(s) to designated parking lots or areas.

Access Control
Because most security incidents (including most covert terrorist attacks) occur inside a building, special attention should be given to controlling building access. The nature and level of access

control (along with visible security measures such as CCTV cameras) also establishes the building's security culture or "image," which is important in deterring criminals in the first place. In cases of small office buildings, management frequently leaves the doors open for tenants and visitors. If the risks are relatively low, this may be acceptable during the office day. Locks on exterior doors that are closed at night and on weekends should be of high-security commercial grade with exterior hinges "pinned."

As an alternative in buildings with only a few tenant employees, general building access might be controlled with each employee having a key, the code to an electronic keypad, or a card operating an electronic card access system. Visitors and delivery persons would have to use a building directory intercom to seek admittance. Depending on the system, tenants would then remotely "buzz" visitors in (convenient, but not very secure), or be required to physically go to the lobby or entry door to admit visitors. Building management would control building deliveries.

Where stricter access control is necessary, buildings may use a security staff (proprietary or contract guard company) to screen tenants and employees. This can be done through use of a building or tenant(s) photo ID card for visual screening by guards, or by means of one electronic card access control system for all. When card access systems are used, employees/tenants can be processed automatically through one or more lines, while visitors can be directed to a special line for screening and search. Temporary (time expiring) badges can be issued to visitors who have been "approved" by tenants, or for access to "public" offices. Electronic card access control systems also have the advantage of keeping track of who's in the building, which is especially important if an evacuation becomes necessary.

Depending on risk, metal detectors to screen all persons entering for guns and knives may be appropriate. X-ray screening of packages, purses and briefcases may also be used. Consideration should be given to requiring that all overnight and courier service pickups and deliveries be directed to a central "mailroom" or desk for appropriate screening. This prevents "delivery" or "courier" persons from roaming the building (and offices) alone and unmonitored.

Note that many large buildings require a combination of technology and manpower to adequately address their security needs. Systems and hardware won't accomplish the entire task, and neither will guards. Integrating both into a comprehensive security plan is required.

Attacking the Building

Terrorist type attacks (this could also include malicious vandalism and major damage by disgruntled building or tenant employees) might be directed against the building itself, rather than just against the tenants and their property. Or a building or tenant employee may use the building itself to facilitate their attack against their employer or against a single tenant. Management should secure utilities (water, power, and gas) which are accessible outside, but on building property. Equal care should be given to securing access to the building ventilation system, including any access points on the individual floors.

Garbage and trash bins and skips are likely locations to hide a bomb. They should be located, whenever possible, at least 100 feet from the building, and be chained or fixed to prevent their being moved back close to the building. The entire exterior base of the building (at least the first story level) should be illuminated with a minimum of one foot-candle of light. Isolated or "risk" areas such as loading or delivery docks should receive special attention, including increased lighting, locking, and observation, to prevent unauthorized access to the building. When a building security staff is available, CCTV may be used to monitor the exterior of the building and any "high risk" areas.

Bomb blasts at nearby buildings (within four or five blocks) may produce bomb damage by blowing out windows at your property, resulting in injury, death, and property damage. While somewhat expensive, blast film over the windows can reduce or prevent injury or damage from shattered flying glass, and can keep glass from falling to the streets below.

Tenant Spaces

Due to changes in modern office building design and operation, many traditional visitor-screening methods, such as elevator operators and office receptionists, have virtually disappeared in many office buildings. Tenants frequently provide access control to their own spaces, sometimes with building security advice and involvement, sometimes on their own. Frequently tenants don't do much screening or control, relying instead on the building's lobby screening (if any).

One advantage of tenants "doing it themselves" is that where a need exists, some tenants can afford to implement higher levels of access control for their space (such as bio-metric access control devices) for their positive identification of their employees. This is often impracticable for large buildings as a whole. Usually this is because of the smaller numbers of employees that have to be accommodated, rather than the thousands tenants and visitors that enter a large office building daily.

Emergency Planning
Every office building, but especially high-rise buildings (higher than seven stories) should have an emergency plan that limits and/or mitigates the impact of any security breach or other disaster. Special attention should be given to developing and practicing building evacuation plans. While evacuation drills are inconvenient in high-rise buildings, they are critical for safety and should be performed at least twice a year. Warning communications are especially important and should be regularly tested (at least every 90 days). Bomb threat assessment, search, and evacuation plans should be included and periodically tested. Building tenants and employees are less likely to panic if they have practiced evacuations and know what to do. And lives will be saved!

Building management may require that its security officers, whether proprietary or contract, be trained in basic first aid and CPR. Security officers are often the first responders to tenant employee or visitor emergencies, especially at night and on weekends.

Periodic Review
Finally, whatever security plan is developed and implemented, it should be periodically reviewed to ensure that it is in fact operating the way it was originally designed, and that it continues adequately to address the threats to the building and its tenants which change over time, sometimes over a very short time. If management does not have the in-house capability to do so, a non-product affiliated security consultant should be retained to assist with the review or audit and provide an independent viewpoint.

Today it is not enough just to be secure and to plan and be prepared for emergencies. In the wake of September 11 building tenants, employees, and visitors are seeking a sense of order and predictability. They must believe that they and their belongings are safe and secure.

Computer Security & Recovery: What Every Business Should Know

By Kevin J. Ripa

Note: This article is extremely technical and unless you are knowledgeable and comfortable with computers, you are advised to hire a technician to do this for you. Please read the disclaimer on the copyright page before you attempt to do this yourself.

In an ordinary world, people recognize their limitations. An ordinary person with no flying experience would never think about hopping into the cockpit of a 747 and flying it across the ocean. This is just common sense. Unfortunately, the computer world is anything BUT ordinary. It is fascinating that we would sit down in front of the most technologically advanced piece of equipment ever devised for personal use, and start using it, completely oblivious to such trivial matters as security, let alone proper use and understanding. Many of us get away with it, suffering from lockups, accidentally deleted files, and the odd virus infection or other security compromise. This does not need to happen, but sadly it does all too frequently. Why? Well here is my theory. The manuals and other training material provided for computers seem to be written at a level that assumes we have a strong understanding of how this silly box works and the terminology that goes with it. Unfortunately, by the time you are able to understand what the book says, you don't need the book anymore. This forces most users to toss the book, forge blindly on, and hope for the best. Well, suffer no more. You are about to learn about this very important topic in language that the rest of us can understand. We will also provide resources and information to pursue the topics further if need be.

Computer security is the most important, if not most ignored part of business computing. The problem does not lie in apathy. Those that know about it are attempting to implement it, but there are many that have no idea what the risks are or what to do to stop them.

Before you start worrying that this will be geared towards large corporations, let me tell you that the complete opposite is true. This information is geared for the rest of us in the setting of a home or single office computer, as well as any small office setup consisting of up to ten computers. Having said that, there are many policies and rules that need to be in place for large networks as well. Let's start with exploring the more common risks that exist.

To spread their product, Virus creators count on millions of users doing something that they would never do in the real world—Open their door to strangers. They prey on users that don't understand how Viruses are spread, inadvertently sending Viruses to everyone they know. By taking a few simple precautions, we can all make accidentally sending and receiving Viruses, a thing of the past.

95% of all Viruses, Worms, and Trojans are written to work within the Windows environment and the Outlook and Outlook Express programs. Why? Because the creator can be sure of doing the most damage. This doesn't mean however, that we should all run out and change our operating systems to Linux or Mac, nor do we need to subscribe to AOL to avoid the largest bulk of malicious emails. Besides, doing that will place you smack in the middle of the largest SPAM machine in existence along with Hotmail and Yahoo! What we need to do is understand the risks that are out there and the methods by which they propagate.

True immunity to Viruses doesn't exist. Just as it's hard to stop a crime before it happens, it's difficult to halt a Virus before it damages at least a few computers. However, just as you can take steps to minimize your risk of becoming a victim of crime, you can minimize your risk of becoming the next victim of either a malicious program or data loss.

There are three main threats within email messages and attachments. They are Viruses, Worms, and Trojans. For the most part, any malicious program/file you receive will fall into one of these three types. Keep in mind that these are not the only threats, but they certainly are the most common.

Viruses—Defined

To be defined as a Virus, a program must replicate itself in order to carry out a mission, be dependent on a "host" to carry out the mission, and create damage to the computer system "infected." A Virus is a program that reproduces itself, hides in other computer code without permission, and does nasty or undesirable things not intended by its victim.

A Virus cannot replicate itself and be sent to others without you telling it to. (even though you do not know you are doing it). In other words, a Virus can only be spread through the users' own intervention.

Viruses are considered a low-level threat because they cannot sneak into your system. They must come either on a disk someone has given you, or through an email with an attachment. An attachment is a file that comes with an email, like a document or a video clip. When you receive email with an attachment using Outlook Express or Outlook, you will see a paper clip in the upper right-hand corner of the email box. When you click on the paper clip, you are shown the name of the attachment and given the option of either opening it or saving it somewhere else on your computer. To carry a Virus the attachment must have an executable file extension like .exe, .com, .vbs or .pif. The only exception to this rule is a Macro virus, which will have a .doc extension. You cannot get a Virus through email by simply reading the email. You must open the attachment to activate it. Beware of email programs that auto-open attachments!

As you can see, you don't just "get" a Virus by accident. You have to let it in. You would not think of opening the door in the middle of the night without first knowing who was there and what they wanted. A Virus can-not hop from computer to computer by itself. You have to send it out.

Worms—Defined
A computer Worm is a reproducing program that runs independently of other programs and travels across network connections (i.e. your Internet connection through email). The main difference between Viruses and Worms is the method by which they reproduce and spread. A Virus depends on a host file, files transferred between machines to spread, while a Worm is completely independent and spreads through network connections. Worms are delivered from within the email message. You do not need an attachment.

Typically, just reading the email can enable the Worm on your system. Because the Worm is a program with its own protection system (the message that fooled you into opening it in the first place) as well as its own executable damage program, it doesn't need any external source, such as an attachment. When a Worm gets into your computer, the program automatically finds the address book in Outlook and/or Outlook Express and sends itself to all your email addresses. You didn't even know it was there. Having said this, there is little point in having a worm by itself. You usually see Viruses attached to them so that when the worm automatically propagates, it takes the Virus with it.

You have two different formats to choose from when sending email: HTML Format or Plain Text Format. (In the case of AOL and

Outlook, you have an enriched plain text option as well.) Worms propagate only in a HTML format environment.

You can receive an email in HTML format that looks no different than an email in Plain Text Format. That's because the author has placed only text in the HTML format email. HTML format is what allows an email to look like a webpage, if the author wishes. Worms would NOT spread if everyone sent their messages in Plain Text Format.

Trojans—Defined

A computer Trojan is a malicious computer program disguised as something useful. The major difference between Viruses and Trojans is that Viruses can be reproduced, while a Trojan is a one-time program, which executes (usually) as soon as it's activated. Trojans are the most common method of gaining unauthorized access to your hard drive. Trojans are considered the highest risk because you can get one and never know it. A Trojan allows intruders to access the Internet, email, and newsgroups through your computer. Any trouble they cause, looks like it came from you. A Trojan can come into your computer attached to an email or embedded in a self-executing link in an HTML email. Plain Text emails cannot carry a Trojan. They must come in an attachment.

More so than ever before, hoaxes abound. Some promise cars from Honda, and some talk about the mother of all viruses. A virus CANNOT work in Windows AND Macintosh AND Linux, etc, etc. This is the first clue to a virus hoax. The text will indicate that the virus will wipe out any of these systems. A virus must be written for a specific program or operating system. It can't cross over. Just RECEIVING an infected attachment CANNOT activate a virus. You must open it for the virus to execute. If you delete it without opening it, you will be safe. Having said that, older versions of Outlook and Outlook Express automatically open attachments in emails unless you have specifically taken steps to stop this. This is why a user should always be updating their programs. A virus CANNOT trigger "the reformat command" within Norton Utilities or McAffee. This is a big scare tactic being used. Norton and McAffee do NOT have reformat commands.

Viruses are the most difficult to stop on their way in, suggesting that you feel the need to open attachments. Do not send attachments unless you have to. If you have typed something in a Word format to send through email, cut and paste the entire article directly into the body of the email. Thus, no attachment. Sometimes we don't have the choice. We need the attachment. At other times we get an attachment and open it without thinking.

Why would you open an attachment from someone you don't know? Or even an unexpected one from someone you know? How hard is it to reply to the message, and ask the sender if he/she sent you an attachment? Or scan it with a good Anti-Virus program? This way, you can warn him/her as well. There is NO replacement for a good Anti-Virus program and some good old-fashioned common sense. We will address both shortly.

One of the most important things you can do—and it has applications all over your computer and not just email—is to enable the showing of file extensions. File extensions are usually three letter monikers (but can be two or four) that follow a file name (i.e. .doc). These extensions tell Windows what type of file it is so that Windows knows what program to use to open it. For example, if the file extension is .jpg, the computer won't try to use Microsoft Word to open it. In order to better explain this, allow me to make you familiar with some of the more common file extensions and what they mean.

- .doc—tells Windows that this is a document text file and to use Microsoft Word to open.
- .exe—this is an executable extension. This means that the file is a PROGRAM. (IMPORTANT)
- .com—once again, an executable program file, but usually at a DOS level.
- .txt—typing of words only.
- .vbs—Visual Basic Script (VERY IMPORTANT)
- .pif—Program Information File.

I have not explained all file extensions because there are over 7000 of them and you will never see most of them. But you should familiarize yourself with the more common ones. If you have forgotten what one means, or if you see one that I haven't defined, you can go to www.webopedia.com/quick_ref/fileextensions.html and look it up. If it is not there, or you don't understand their definition, simply go to the Internet and put the extension into your search engine. I highly recommend using www.google.com for a search engine.

You may be asking yourself, "What have these got to do with me?" Let me explain. Now that you know that .doc is a Word file like a report, when your co-worker sends you something that is supposed to be a report and you see that the extension is .exe, you should know that something is wrong because .exe is a program execution file. Many of you are looking at your computer and thinking, "I don't see these extensions of which he speaks." All you see is the report you typed named "The Wonderful World of Widgets." All your files look like this because you don't have the showing of extensions enabled. Once you do, that file will look like this—"Wonderful World of Widgets.doc." Now you can easily identify it by the .doc telling you that it is a document. First I will explain why this is important and then I will explain how to change your settings to show them. Having said that, remember that even a trusted incoming .doc file should be scanned because it could have a Macros Virus embedded in it from the sender's computer who isn't even aware of the infection!

.vbs is the file extension for Visual Basic Script. The ILOVEYOU and NIMDA Viruses, among others, had this extension. It needed the program on Windows that opens .vbs files in order to operate. With the first permutation of the ILOVEYOU Virus that came out, it was named ILOVEYOU.txt.vbs.

Unfortunately, this was done purposely under the assumption that most people don't have file extension viewing enabled. Therefore, all you saw was ILOVEYOU.txt. The .txt was not an extension, but rather it was a part of the title that was typed that way to fool you into thinking that it was a .txt file. It was geared towards people who know just enough about computers to be wary of receiving an unsolicited file but had the idea that a .txt file was merely a text file, and so therefore not very dangerous. It worked. If you had showing of file extensions enabled, you would have seen the .vbs and even though you may not know what .vbs stands for, you might have thought it odd that there were what appeared to be two file extensions. This seems to be the M.O. of many Virus writers now. At no time did you stop to think that you didn't have viewing of extensions disabled, so why would you see the .txt extension at all? Do you see the importance of file extensions?

To show file extensions:

XP—Luna

1. Select Start\All Programs\Accessories\Windows Explorer.
2. Select Tools\Folder Options. The Folder Options dialog box appears.
3. Select the View tab and locate the "Hide extensions for known file types" option.
4. Deselect the "Hide extensions for known file types" option.
5. Select OK.

XP—Classic

1. Select Start\Programs\Accessories\Windows Explorer.

2. Select Tools\Folder Options. The Folder Options dialog box appears.
3. Select the View Tab and locate the "Hide extensions for known file types" option.
4. Select OK.

2000 Pro, ME

1. Select Start\Programs\Windows Explorer. The Windows Explorer appears.
2. Select View\Folder Options. The Folder Options dialog box appears.
3. Select the View tab and locate the "Hide file extensions for known type files."
4. Click the check mark to remove it.
5. Click OK.

Win 98, 98SE

1. Select Start\Programs\Windows Explorer. The Windows Explorer appears.
2. Select View\Folder Options. The Folder Options dialog box appears.
3. Select the View tab and locate the "Hide file extensions for known type files."
4. Click the check box to remove the check.

Now, any files that are anywhere on your computer will have extensions after the names. Also, when you get an attachment to an email, you will be able to see by clicking the paper clip, the FULL name of the file. So if you get a file that is supposed to be a .doc file and you see .doc.exe at the end, you know something is wrong.

There are two very important things you need to do to avoid getting ANY .vbs worm/Virus combination. First, you can click on Start/Windows Update, and it will take you to a website where you download all of the Microsoft bug patches and security fixes as they come out. You can also download from there a Critical Update Notification. This will automatically inform you when there are new fixes of ANY type for Windows operating systems. If you do this religiously, you won't have a problem. The patch for the security hole exploited by .vbs script was out about three weeks before the actual ILOVEYOU Virus came out.

Whenever you double-click on a Windows file, an action associated with the file's format occurs. The default action for double clicking on a Visual Basic Script file—for example, the NIMDA virus—is to execute, or open, the script contained in the file. This is bad, which brings us to the second thing we must do. We must change the default action to Edit, which causes the file to open in Notepad rather than to execute the script.

XP—Luna

1. Click Start/All Programs/Accessories/Windows Explorer.
2. Select Tools\Folder Options.
3. Select the File Types tab and locate the VBScript File. All the files are in alphabetical order.
4. Click the Advanced button. The Edit File Type dialog box appears.
5. Select the Set Default button.
6. Click OK to close the Edit File Type dialog box.
7. Click Close to close the Folder Options dialog box.

XP—Classic

Select Start\Programs\Accessories\Windows Explorer.
1. Select Tools\Folder Options.
2. Select the File Types tab and locate the VB Script file.
3. Click the Advanced button. The Edit File Type dialog box appears.
4. Select the Set Default button.
5. Click OK to close the Edit File Type dialog box.
6. Click Close to close the Folder Options dialog box.

2000, ME

1. Click Start\Programs\Accessories\Windows Explorer.
2. Select Tools\Folder Options. The Folder Options dialog box appears.
3. Select the File Types tab and scroll to VBScript Script File. Note: If there are two VB Script files, you will have to repeat this procedure for each one.
4. Click the Advanced button. The Edit File Type dialog box appears.
5. Select Edit.
6. Click Set Default button.
7. Click OK to close the Edit File Type dialog box.
8. Click Close to close the Folder Options dialog box.

Win 98, 98SE

1. Select Start\Programs\Windows Explorer.
2. Select View Menu\Folder Options.

3. Select the File Types tab and scroll to VBScript Script File. Note: If there are two, you will have to repeat this procedure for each one.
4. Select the Edit button. The Edit File Type dialog box appears.
5. Select Edit and click OK.
6. Click OK to close the Folder Options dialog box.

In some older systems the Edit function may not appear. In such instances, click the New button and enter Edit in the action field and Notepad.exe in the application field. Once Edit appears, make it the default action as shown above.

If you follow these steps and don't see the VBScript Script File, this means that your Visual Basic Script Program is not installed from your operating system disk. This should only happen in the first edition of Windows 98. If you have two instances of the VBScript Script File, change them both. Beyond all of this, there is no substitute for a good anti-virus program.

I explained earlier how you get worms and what they do, and we covered some virus and worm prevention in the last section. So what happens if, despite all of your best intentions, you still get infected with a worm? Well, you can stop them from sending themselves out to everyone in your address book, which is what worms do. To avoid sending Worms, set your email to send in Plain Text Only.

To set the email format to Plain Text Only in Outlook Express:

1. Click on Tools/Options/Send. At the bottom of that window, you will see Mail Sending Format and News Sending Format.
2. Set both to Plain Text.
3. Click Apply.

To set the email format to Plain Text Only in Outlook:

1. Click on Tools/Options/Mail Format.
2. Set to Plain Text.
3. Click Apply.

Now there is no way to send the Worm out of your computer because the HTML it needs to propagate is now disabled.

I don't know of any way of protecting against Trojans other than a good Firewall, a good anti-virus program, and good old-fashioned common sense. (Do you see a pattern here?) Use the same diligence to protect against these that you would for viruses inside attachments.

The next step to take to avoid malicious programming is to have a good anti-virus program. I am a staunch supporter of Norton Anti-Virus for reasons too lengthy and numerous to cover here. Remember, though, an anti-virus program is virtually useless unless you update it frequently. I have seen numerous computers "protected" by an anti-virus program and the user can't understand why they got a virus. I ask them when they last updated their program and they usually answer, "Update? What do you mean?" Fortunately, most anti-virus software provides these updates free. I recommend updating at least once per week. Given that there are an average of three new viruses released every DAY, staying updated is extremely important. In the case of Norton Anti-Virus, as soon as a new update is available, you are informed about it.

In the unlikely event that you ever get a virus, have no fear. There are any number of people who can be contacted, myself included, to assist in the removal. In the vast majority of cases, this can be done over the telephone. Do NOT get mislead by so called "virus removal experts" who do little more than run little programs on your computer that they have taken from the anti-virus websites. These DO NOT always remove the viruses. They merely stop them from continuing the damage they were intended to do. Proper virus removal usually has to be done line by line to ensure it is truly removed.

Now that we have secured our computers against malicious programming, let us take a look at some trouble that an anti-virus program won't protect you from, and that is access to your computers from an unauthorized user, otherwise known as hacking.

Hackers
Computer hackers break into systems for various reasons. These can range from showing off, to causing damage, and even stealing CPU time. What would they do with stolen CPU time? The industry standard at this time for encryption is 128 bits. This is the minimum that a website must prove before it is allowed to accept credit cards. In theory, it would take the fastest computer in the world far too long to crack something encrypted with 128 bit for it to be useful to steal, say, credit card numbers. That is one computer. Imagine if I could feed portions of that same code into 1000 computers and have them simultaneously working on cracking the code.

Having your computer protected by a password is not nearly secure enough, although it is a necessity. Passwords only keep the honest people out. Once a hacker is inside your system, if he is met with a prompt for a password, he will use "Brute Force" programs that methodically enter all combinations of letters, numbers and characters until they stumble on the right one. Although some of the more rudimentary programs can do upwards of 30,000 combinations per second, longer passwords obviously still take a long time to crack. For a 6-letter password that is lowercase and only letters, this equals 308,915,776 possible combinations, taking approximately three hours to crack. For an 8-letter password that is lowercase and only letters, this equals 208,827,064,576 possible combinations, taking approximately 1933 hours, or 80.5 days to crack. See what a difference two characters make?

Now imagine adding other variables such as uppercase letters, numbers, and characters into the fray, and it isn't hard to see how to set up a password. A minimum 8-character password with a combination of uppercase and lowercase letters, as well as numbers becomes a formidable defense. Do this with both username and password, and the difficulty doubles. The important thing to remember is that you don't want a password that is simply a word that you can remember. Imagine the word JUSTIFICATION as our password. It is 13 characters long, so you would think it would be secure. Most brute force programs first run dictionary lists before they start with random character scans. In other words, they run every word in the dictionary against your computer password. Webster's has approximately 300,000 words in it, of which JUSTIFICATION is one. At 30,000 combinations per second, it would take less than 10 seconds to break that password. Here are my recommendations for selecting a password:

- Your password should be at least eight characters.
- Your password should contain a combination of upper and lowercase letters as well as numbers and punctuation symbols.
- Your password should NOT contain any part of your username.
- You should change your password every 30 days.
- When you create a new password, it should not contain any part of the previous password.
- Your password should not contain a common word or name.
- You should select a password whose characters you can easily remember.

Everyone is now throwing their book out the window, screaming about how they will never remember their password easily if they are meeting the above requirements. Let me show you two examples.

Example 1

Mh1llifwwas!

The above password is 12 characters long. Well beyond our minimum of 8. It has upper case, lower case, a number, and a punctuation symbol. You ask yourself, "How in the heck do I remember that?" My answer is easily. Here is how.

Mary had 1 little lamb its fleece was white as snow!
The password was made from the first letter of each word in this well known nursery rhyme.

Example 2

K1v2n-R3p4

Again, more than our 8-character minimum, with uppercase, lowercase, numbers, and symbols. But what does it mean? If you bothered to read far enough down the cover of the book, you would have seen that my name is Kevin Ripa. I have merely taken out the vowels and replaced them with consecutive numbers. Add a dash between the names or a space works too!

All it takes is a little ingenuity. Be creative and be secure. Oh, by the way, don't use the same password and username for everything.

When a hacker breaks into a system, the hacker normally does not have an unlimited amount of time to peruse the system's files. Depending on the system type, the hacker will start his or her search in specific locations. For example, within a Windows-based environment, the hacker will quite likely start searching within the MY DOCUMENTS folder. By simply moving your files from this folder, you will make it more difficult for a hacker to locate the information he or she desires. As it turns out, many viruses also hunt for files that reside in specific folders (again, the MY DOCUMENTS folder is a primary target). By moving files or by renaming such folders, you may also reduce your system's risk of a virus attack.

So how do we protect ourselves from this? The program you want is called a firewall. It is software that installs on your computer and blocks access to unauthorized persons. There are a number of different firewalls on the market geared toward personal computer consumption. My recommendation is Zone Alarm, available at www.zonelabs.com. Believe it or not, it is a free download for personal use, and very reasonably priced for business use. When properly configured, this program will not only stop external access, but it will make you invisible to outsiders scanning a range of addresses that your computer would fall under. With small offices running networks of more than three computers that are all hooked to the Internet, I would highly recommend getting a Router. Space prohibits me from getting into detail, but if you talk to a trusted computer technician, they can explain it very well.

If you would like to test your computer and see how vulnerable you are to outside attack, go to www.grc.com, and run the Shields Up! test, as well as the Leak test. As if the above examples of computer risks weren't enough, here are some more to tempt your cyber taste buds.

- War Dialers—Anyone who has a telephone line connected to their computer is at risk of this. Hackers use a program that continuously dials numbers in sequence. When the number is answered, the program determines whether the answering number is a voice line, fax line, or computer line. If it determines that the answering line is a computer line, it logs it for the user's future reference. Very popular for vendetta-type attacks.
- Vulnerability Scanners—A program that scans a system for all open ports and logs the vulnerabilities for the user's later reference. It is seen more often for the placing of Trojans or access to a computer as a launch point for something else.
- Computer Microphones—If a hacker gains access to your computer, he can use your microphone to listen to everything in the room your computer is situated. Very popular for corporate espionage.
- Ping of Death—Sending a large amount of information to your computer in a Ping that causes your computer to crash.
- Packet Storm—Sending many packets continuously to a computer to completely tie up its resources and lock it up or crash it.

Not worried yet? Here are a few others. IP Fragmentation, SYNflood, Connection, Kill Session, DNS Spoofing, Insertion and Replay,

Delay, Degradation, Simple Direct, Simple Indirect, Progressive, Distributed, Attack by Combination, Fragmented, Deception, Tunneling, Immobilization, Timing, Attack on Prep, Second Intent, Counter Time, and Broken Rhythm.

A good firewall will stop most of this. Even if a war dialer—a hacker who tries to use your computer for vendetta-type attacks—connection succeeds, Zone Alarm will make sure that no information is taken out of the computer.

How Do YOU Look From the Internet?

Many people, not knowing any difference, enter their real name and company name into the computer's run-once-only program at startup when they buy a new computer. Believe it or not, this is visible to anyone on the Internet if they know how to read it. If you visit my website, I can see that information. As a matter of fact, when using a cable modem for high speed Internet, it can be seen in a simple IP trace! Not sure what your computer's name is? On your desktop, right click on the My Computer icon and select Properties. When the box opens, you will see what your computer name is under the "Registered To" section. Here is how you change your computer's name to something less identifiable.

*******Be VERY careful and follow these instructions TO THE LETTER! You can cause serious damage to your computer if you click in the wrong place, so read twice, click once!*******

Click Start, then Run. Type in "regedit" without the quotes and click OK. Now you are into the central nervous system of your computer.

ONE WRONG CLICK AND YOU CAN PARALYZE IT!

BEWARE THAT THERE MAY BE FOLDERS THAT LOOK VERY CLOSE IN NAME TO THE FOLDERS I REFER TO. READ CAREFULLY AND BE SURE TO ONLY REFERENCE THE FOLDER I LIST EXACTLY!!!!

1. Click on the plus sign beside HKEY_LOCAL_MACHINE. *A new list will drop down.*
2. Scroll down the new list and click on the plus sign beside SOFTWARE. *A new list will drop down.*
3. Scroll down the new list and click on the plus sign beside MICROSOFT. *A new list will drop down.*
4. Scroll down the new list and click on the plus sign beside WINDOWS. (WINDOWSNT for NT and 2000 Pro users). *A new list will drop down.*
5. Scroll down the new list and click on THE FOLDER ITSELF named CURRENT VERSION.

You will see all of the contents of this folder on the right hand side of the screen. This is all identifying info about your computer in various capacities. Scroll down the list on the right side until you come to the Name REGISTERED ORGANIZATION and/or REGISTERED OWNER. The procedure for changing both is the same, so I will only explain it for changing registered owner name. Then you can repeat for changing registered organization if you wish.

1. Click TWICE on the words Registered Owner.
2. A box will open up called Edit String.
3. As you can see in the Value Data box, your name is listed.
4. Select the data in there, i.e. Your Name and delete it.
5. Type in a new name for your computer. *This can be anything you want. I use Default for both.*
6. Click OK.

CLOSE THE REGISTRY EDITOR BY CLICKING ON THE x BOX IN THE TOP RIGHT HAND CORNER OF THE WINDOW.

Now the name is changed. To see the name change reflected, right click on the My Computer icon on your desktop. Select Properties. Make sure you are looking at the General tab, and there it is.

OOPS! I Did it Again!

You have suffered a system crash in the middle of typing that long report and you didn't save it. Now you have to start typing all over again. You clicked the wrong button and accidentally deleted a file or photos. Your computer just won't start and keeps saying your hard drive can't be found. Any of this sound familiar? In many cases the people who suffer these tragedies don't realize that nothing on a computer is ever truly lost. What you need is a data recovery specialist. Data Recovery is the fine art of recovering lost information on a computer, including deleted files, crashed hard drives, and virtually all information you may have thought was gone forever. This can also be done even if your hard drive has been reformatted! Rest assured it is not very cheap, but explore the alternatives. We recently recovered an entire hard drive that had been reformatted on advice from a tech support person. The customer didn't realize that a reformat meant that all her data would be lost. When the reformat was complete, she realized that three years worth of work was gone. We were able to recover EVERYTHING.

Let us take this one step further. Beyond data recovery, there exists a world called Forensic Data Recovery. This would be the extraction of files and information from a computer that a criminal may have been using. It includes data that may have been stolen by a disgruntled employee. It includes finding hidden or deleted files, including emails, that may prove a less than scrupulous employee. Forensic data recovery involves recovering data from a computer (basically) without turning it on. When a computer is turned on, it changes various settings. The change in these settings is enough to render the evidence on the computer unusable. Some of the things we have seen include hiding information on hidden clusters, embedding information invisibly in a document, renaming extensions to bury the document, etc. We have also seen an instance where a hacker illegally accessed someone's computer and was storing child pornography on it. By doing this, the hacker never had to worry about getting caught. His computer was always clean. The unsuspecting victim was arrested and charged with possession of child pornography. He lost his wife, kids, house, job, everything. His defense attorney hired us and we were able to forensically analyze the hard drive and show that the images had been placed there remotely and accessed numerous times when the victim couldn't have possibly been near his computer.

It is my sincere hope that these instructions are presented in a common sense, easy to understand manner. I hope that you can see that computer security is not the magical, mystical beast that many would have you believe. Computer security can be exercised by anyone just following a few simple steps once a week. Should you ever have any questions, you can contact me by email or visit my website at www.ComputerEvidenceRecovery.com. It is time to take over your computer and stop letting it and malicious programming, rule you!

Kevin J. Ripa, owner of Computer Evidence Recovery, Inc., is a former member of the Department of National Defense serving in both foreign and domestic postings. After he retired from the federal government in 1991, he continued to serve on an "on call" basis. Mr. Ripa provides services to the legal community, insurance industry, civil government, law enforcement, corporate entities, and private citizens in Alberta and around the world. He is respected and sought after by the investigative community throughout North America for his expertise in Computer Investigations.

Security and Company Culture
By Michael G. McCourt

The tragic events of September 11 and subsequent anthrax attacks forever changed the landscape of security in our country.

Before September 11 we thought of ourselves as invincible. Now, we recognize our vulnerability. Our nation has scrambled to create an Office of Homeland Security, called upon our National Guard to provide support, increased the visibility and budgets of law enforcement agencies across the country, and devoted resources to developing medical remedies for diseases thought to be extinct.

Corporations, too, have joined the race to develop disaster plans, increase security technology, add security personnel, develop policies, and rethink their allocation of resources. These "quick fix" strategies have cost billions of dollars. Despite the phenomenal increase in the cost of safety and security, industry reports say the general public (and employees in particular) does not feel any safer. Why hasn't the increase translated into a greater sense of security?

The answer, simple as it is, has been overlooked for decades in all but the largest of corporations. In order for security to be successful, it has to be a cost-effective function that is woven into the fabric or culture of the corporation.

The first element, cost, is easily defined but challenging nonetheless. Companies allocate financial resources based on return-on-investment (ROI). A declining economy prior to September 11, along with the catastrophic events of that day, has focused more attention than usual on ROI issues.

ROI utilizes a set of varying metrics to measure success. The measure of success in security is "nothingness." (Nothing happened yesterday, nothing is happening today and, hopefully, nothing will happen tomorrow.) Corporate leaders do not as easily embrace this "nothingness" concept, widely accepted by security professionals. It is critically important for senior management to become more aware of, and comfortable with, this metric.

To be effective, security requires a budget equal to, or slightly greater than, the level of threat faced by the organization. Depending on the industry, that number can be significant. Companies have to resign themselves to the fact that security comes at a price. Negligent security, however, comes at a much higher price.

Research indicates that the average out-of-court settlement for a negligent security case is in excess of $500,000, with the average jury award exceeding $1.5 million. The insurance axiom—"it's not a question of whether you can afford it, it's a question of whether you can afford not to have it"—applies to security as well.

At a time when the future is less predictable than ever before, the question of whether or not an organization can afford security should be a nonnegotiable, easy answer. Corporations will spend significant amounts of money on programs and technology viewed as being intrinsically tied to the bottom line. Until the time comes when security is viewed in the same regard, companies will continue to expose themselves to risk by providing substandard security programs. However, money alone is not the answer.

Dollars aside, the real challenge for organizations is to integrate security and corporate culture. To marry the two effectively, leadership must be able to define their organizational culture in concrete terms, not by reciting a mission statement or a pie-in-the-sky list of attributes, which may or may not be practiced on a daily basis. Although culture includes many of those elements, it tends to be more oblique and far-reaching in its impact on daily operations.

Every organization has a culture, whether or not it is purposefully defined. Culture is the glue, or set of unspoken guidelines, that dictates employee behavior, guides the decision-making process, influences discipline, supports (or fails to support) customer service, and ultimately impacts the organization's productivity and profitability. It's not found in any corporate document and is rarely discussed in boardrooms, but its impact is felt in every interaction, at every level of the organization, every day.

What, then, determines corporate culture? Generally, the CEO and the senior staff of an organization define culture. It is based on their personal beliefs, behaviors, operating standards, and core ethics. If personal safety, respectful and open communication, honesty in reporting, and a belief that every person has value and contributes to the success of the organization are among the leadership's core values, then safety and security may flourish. If, on the other hand, the values listed above are not seen as critical to the success of the organization, then safety and security will take a back seat to programs more directly related to the technical and operational side of the business.

Organizations that have successfully woven safety and security into their culture will factor security into every project, every decision, and every action at every level of the organization. Without the support of senior management, security is doomed to become a part-time program, subject to the "issue of the month" methodology of management.

What can an organization do to integrate security into its culture? Adopt the following strategies:

- Create the position of Chief Safety Officer (CSO) and grant that individual access to the board of directors.
- Ensure that individual's assigned responsibility for security will receive adequate training.
- Make security an agenda item at every board or senior staff meeting. Include articles on safety and security in corporate newsletters.
- Develop reward incentives for employees reporting safety and security breaches.
- Sponsor safety and security-related programs within your community.
- Develop disaster plans capable of addressing today's unique environment.
- Post signs and information relating to safety and security throughout the organization.
- Develop policies addressing safety and security and enforce them fairly and consistently.

Often, solid security strategies will be objected to on the premise that they will cost too much money. Ironically (as these points demonstrate), successful security costs more time than money.

The time to integrate security into corporate culture is now. Corporate leaders have a choice to make: they can continue to underestimate the importance of security and pay the price down the line, or they can make a conscious decision to incorporate security into their culture and defend their human resources, capital investments, and intellectual properties against the challenges we face in the 21st century.

"Reproduced with permission from STP Specialty Technical Publishers." Michael G. McCourt is the president of Michael G. McCourt Associates, Inc., a Massachusetts based consulting firm specializing in workplace violence prevention and counseling.

Developing Approaches To Business Security
By David P. Roberts

Before delving into what we can expect to achieve for the future, we must review the historic and current situations, and how U.S. commerce has approached and is reacting to those situations.

We need to do this so we can build a foundation upon which the futuristic security-oriented superstructure of corporate America can be created, developed, and become the common expectation.

It would be fair to comment that from a personal and professional perspective the prevailing attitudes I have encountered are both salutary and disappointing.

A number of prevailing attitudes that give rise to that commentary can be delineated as follows:

Commerce appears to be more concerned about the "Security of Profits" than they are about the "Security of Personnel." Few, if any, appear to have understood the concept that to make your business secure, you need to make the personnel responsible for sustaining that business and generating the all-important profits secure in their given environments. Look after your staff, and they will look after the company.

In all but the more far-reaching and innovative thinking organizations, the dedication of funds to security applications is considered a "lost-lead." It is an expense, and a budgetary stigma that immediately after it is begun, is subjected to cost-cutting tactics, the emphasis being on achieving a "bare necessities" status and creating a "bare-bones" operation that hardly meets, and seldom exceeds, basic corporate legislative dictates.

The principle of "pay peanuts and get monkeys" prevails to this day, despite the glaring lessons to be drawn from the World Trade Center debacle.

Be in no doubt that the so-called enhancement of security procedures at airports and other commercial entities, entailing the throwing of more meat at the problem, is nothing short of "ineffective cosmetology," with but a minor application towards the creation of a deterrent effect. When the former Chief of Security of El-Al Airlines reports that he does not feel safe on any U.S. airline, there appears to be little need for stating what to most security specialists is patently obvious with respect to the current level of and application towards the safety and security of the U.S. airline industry.

The displaced emphasis regarding the overwhelming failings of the corporate world to identify with the seriousness of security per se can perhaps best be illustrated by recognizing the identity of those responsible for security applications within their individual corporate infrastructures.

In far too many cases the responsibility falls on the shoulders of premises and facilities managers—those individuals indirectly responsible for ensuring that the latrines contain adequate supplies of toilet paper and soap, among a host of similarly inconsequential banalities and nuances of institutional corporate protocol.

(Please excuse the frivolous correlation—it is designed purely to make a point. One does not expect a heart surgeon specialist to psychoanalyze a diseased mind—he may have a better than average knowledge, but he is not a specialist in that department. I would respectfully contend the task of security should be an exclusive and dedicated role within the U.S. corporate infrastructure.)

Though highly qualified in their own right, I have yet to encounter a facilities or premises manager who has either appropriate global security skills (beyond a basic or generalized concept of access control) or the time and/or opportunity to apply whatever skills they have acquired to enhance their individual security knowledge "along the way" in the course of their respective corporate careers.

Such experience is generally gained in circumstances where their terms of reference are so otherwise extensive and, by definition, adversely prioritized as far as the application of security is concerned.

I have yet to attend a symposium where an engineer, architect, and/or head of a corporate planning and development department has ever introduced a product, facility, or asset that places security as a priority issue.

I have not heard anyone consider and contemplate the destructive attributes associated with a bomb blast! Few people realize that the concussion from the explosion actually "bounces!" So, where one may be well hidden behind a solid structure and out of view of the explosion, the full affect of that explosion could annihilate them just as completely as if they had been standing in direct line of sight to the explosion.

I have heard no one consider the advantages to be associated with the placing of explosion-retaining window dressings over their office windows, or applying a covering to those windows to dissipate, if not obviate, the extremely debilitating effects of flying shards of glass, the effects of which can be the equivalent of forcing a human body through a commercial shredder.

I have heard no one mention or suggest any application that deals with the effect of an explosion from the perspective of the ensuing implosion that always accompanies every explosion, particularly within an enclosed space such as an office environment. The resulting effect is therefore that of "double jeopardy!"

I am aware that the Pentagon was in the process of strengthening the external walls of that building by the insertion of a steel supportive framework for the external wall cavities. I am also aware that they were installing bombproof windows in situ. Indeed, it is in deference to this style of construction that many more lives were saved when the aircraft slammed into that building on September 11.

But, at $10,000 a window, I suspect that corporate planners and developers will not readily suggest such an installation in any new building for the immediate future.

I completed two commercial security audits at premises within the immediate proximity of the World Trade Center. The last one was completed on September 4, with one of the buildings directly opposite the WTC towers at One Liberty Plaza.

In both instances I was obliged to castigate the incumbents responsible for corporate security applications. One was a former security guard with whom the corporation was "comfortable" with regard to his long-term commitment to and familiarity with the company and its personnel. The other was a consortium of departmental premises and facilities managers who, despite possessing bags of common sense, knew little about the intricacies of security beyond access control. Among other obvious failings they individually and/or collectively had not come close to contemplating a plan of action to deal with the sort of terrorist attack that soon manifested itself at the World Trade Center. Even when such apparent deficiencies were identified, there were further elements of frustration to encounter.

In the corporate world, particularly in the U.S., there is a prevailing arrogance to overcome. No one truly believed that any such individual or group could manifest his evil to such devastating effect in the veritable financial nerve center of the world. We were wrong!

In the event that a security expert, engaged to perform a security audit, identifies a problem that is not thereafter addressed and that ultimately results in injury, the loss of life, and/or damage to property, and in the event that such failings existed, were a matter-of-record, and were not acted upon exposes the corporation concerned to the real potential for litigation.

It matters not whether an external or internal security specialist identified the problem as long as there was a documented record, and there is no solace or comfort to be drawn from the fact that such failings were identified internally and to the exclusion of external expertise.

There is no shortage of whistle-blowers protected by law who would be more than willing to point the appropriate finger, more especially if there is a potential for personal criticism involved. And, based on our current corporate dogma, we all need someone to blame.

After September 11 arrogance has been displaced with a palpable element of humility, coupled with both recognition and acceptance of inbred failings to contemplate what, to specialists in the field, was inevitable.

However, with the emphasis on security of profits over personnel and the risk of culpability for negative reaction to professional recommendations, the results manifest themselves in the prevailing threat of litigation winning hands down over the need to address security-related issues.

There is no solace or self-satisfaction derived from not having to state the equally inevitable: "I told you so." Of the cited examples I refer to, it was unnecessary and unwarranted. Some very hard lessons had been learned.

The Making of A Terrorist in the Middle East
In moving on from my personal perception of what ails corporate America, we must seek to understand what it is that forms the makeup and very essence of a modern-day terrorist before we can move to counter the ongoing threat that now exists and will surely increase in the years ahead.

In the Middle East perceptions and reality have become so blended that they are one. This process of constantly repeated flimflamming has deceived terrorists and counter-terrorists alike who do not always act or react based on facts, but only on how they "perceive" the facts.

"Self-evident" truths are only what they want them to be and consequently, the timeworn classical circumstance conducive to breeding terrorist violence continues in the Middle East (as elsewhere).

Those "self-evident" truths fall into the following categories:

- Self-consciousness - Segregated ethnic, cultural or religious minorities - Economic depression - Political Oppression
- External encouragement and/or assistance (e.g., arms, terrorist training, etc.)
- Historical "they are to blame" scenarios - Societies with little history of true democracy.

Add to these the ingredient of Arab nationalism, driven within an Islamic framework that has politicized religion, wherein the faithful believe it is their unique secular mission to unite the Arab people into an Arab Islamic nation of the future. As a result, you have "followers" who now constitute a core of martyrs who envision themselves as the spark to revive their classic Islamic civilization.

Although it has been said, "terrorism is not a product of television or the media," the impressions made by the camera have trapped the current day terrorist realities with stereotyped illusions, sufficient to create a pool of recruitable, potential terrorists seeking recognition, Islamic fulfillment, and revenge.

This situation is naturally exacerbated by the tendency of the U.S. media to "create" the news, rather than report it. Their individual station ratings depend on it!

Yes, terrorist attacks have become bloodier and resultant death tolls have substantially increased, despite what some western counter-terrorist analysts have concluded as major "victories" over terrorist groups by America and its allies.

The recent and most often cited example in the Middle East is the Israeli "smashing" of the Palestinian Liberation Organization (PLO) during its 1982 invasion of Lebanon. However, that situation that continues with even more fervor today than it did in 1982.

Yet, paradoxically, some of these same analysts remain increaseingly worried by the jump in indiscriminate bombings with multiple deaths perpetuated by not only the PLO, but by new terrorist groups, such as the Iranian backed Islamic Jihad, Hamas, and others. Ironically, despite 25 years and more of reported counter-terrorist successes, terrorist violence increases.

Tangible victories (in Western terms) by terrorists had been negligible prior to September 11. Terrorists have found little difficulty recruiting new candidates for death.

Classical circumstances and conditions conducive to terrorist violence continue in the Middle East with no abatement in sight. On the contrary, there is every reason to anticipate and expect an escalation on a worldwide basis in general and here in the U.S. in particular.

Classical counter-terrorist tactics such as Israeli "Iron Fist" operations continue to provide "victories," but paradoxically this technique has not arrested the recruitment of new terrorist candidates. Indeed, as has been seen of late, it breeds still more terrorism and more eager martyrs.

Analytical obsessions with our definition of "victory" over Middle East Islamic terrorists may cloud our ability to understand their persistence and the long-term commitment of the Muslim mind-set.

This "western-induced" blind spot may cause errors in our judgment in the allocation of resources—human, financial, material, and moral—against this threat.

Terrorism and counter-terrorism disciplines and principles not only involve armed violence and subsequent causalities, but they involve the battle for hearts and minds as well. It's psychological warfare: "Get world opinions to condemn and turn against one side or the other."

For example: We currently do not have the support of many Middle Eastern allies, such as prevailed during the Gulf War in the early 90's.

The theory of relativism applies to "Western," not "Islamic," definitions of "victory." What Western culture sees as a defeat for Middle East terrorists may be viewed by those terrorists as a victory. Thus, how can the West ever win while we continue to allow this perception to prevail?

Politicized religion exploits the Islamic psyche and encourages it to transcend death by finding a meaning for one's life. He (the terrorist) wants to count and does not fear death itself, but only extinction with insignificance. Consequently, we have seen an increase in the terrorist martyrs and the trend will continue.

Western corporate and governmental security professionals responsible for counter-terrorist programs need to be perceptive beyond the mere mechanics of crisis-management counter-measures.

Motivation factors, political and socio-economic, as well as the significant influence of Islam need to be more fully understood in developing effective counter-terrorist programs in the Middle East and against their potential targets worldwide.

This situation should bring an urgent need of application into the crisis management thinking of every security professional in every Western corporation! Modern terrorism persists as a direct conesquence of today's worldwide political and economic environment.

It is a growing and increasingly costly threat to the United States and other Western nations; never more so than in these times when the world becomes smaller and the push of corporate America continues to transcend boundaries and borders that were hitherto unapproachable.

For politically and militarily weak revolutionary groups, sub-revolutionary ethnic societies, or antiestablishment anarchists, terrorism is an inexpensive means for promoting their cause by challenging legitimate government authority. Terrorists mean to exploit the fragile yet total dependency of Western civilization upon modern technology. They have now reached the full potential for inflicting mass destruction and disruption.

The threat is real, it is dangerous, it will continue, and as suggested earlier, it cannot be foiled by cosmetic applications and the current tendency towards short memories in the hope that the reality of the threat will dissipate and simply go away. Ladies and gentlemen, that is not going to happen!

Terrorists have the advantage of surprise in perpetrating any incident. They pick the time, place, methodology, target, weapons, and participants. In the case of the U.S. counterintelligence, efforts to negate the surprise factor by penetrating terrorist groups have been stymied because:

- Terrorist organizations "carry no passengers," and generally commit members to acts of violence in the name of "the cause" as proof of reliability before accepting them into a compartmentalized cellular attack unit.
- U.S. counterintelligence morality and law (until recently in any event) specifically prohibited U.S. penetration agents from conspiring to or becoming involved in such acts as assassination. Under these circumstances, of course, direct U.S. intelligence agent penetration of terrorist groups was proscribed.

A potential answer to the problem is selectively to expand U.S. corporate intelligence information capabilities, in conjunction with government agencies, and assimilate these additional areas of relevant data reconstructing dual-track scenarios of terrorist incidents.

Thus, not only are events historically recounted and analyzed in context, but also trigger mechanisms, target selection processes, and psychological profiles are identified, revealing patterns that could potentially help minimize the terrorist's advantage of surprise against corporate America in general and your corporation in particular.

Background—Problems & Solutions
There are no absolutes to fit all causes. The varieties of terrorism that exist are both being deployed and are also yet to fully manifest themselves.

Problems now being reported relate to a form of "internal terrorism". Recent events have resulted in many employee displacements and firings. Reports of embittered employees sabotaging corporate computer networks now abound.

But there is consensual support for several aspects of the problem:

- Terrorists threaten to use and do extraordinary violence.
- Terrorist objectives go beyond immediate destruction, death or injury to selected targets.
- Terrorists select targets for symbolism rather than intrinsic value.
- Terrorist acts are designed to make an impact beyond targets of immediate incidents.
- Terrorists are "actors in the world theater", playing a deadly game of intimidation and fear, with "captive audiences" always available, in hope of gaining support and sympathy for their causes.

To be effective, corporate counter-terrorism must correctly identify terrorist objectives beyond the symbolic, and attempt to make safe other potential targets by thinking outside the proverbial, currently well-established, box.

Information collection and accurate, perceptive intelligence interpretation and analysis are the key to successful anti-terrorist action at both governmental and corporate security departmental level.

The ultimate purpose for integrating collected terrorist information into intelligence estimates is to pre-empt future terrorist events and thereby negate the surprise factor. Realistically, can it be done? To a limited degree, yes. However, there are far too many variables to make predictability absolute. But there are discernable patterns that need to be more closely studied and applied. Additional new questions need to be asked, answered, and added to reconstructed historical scenarios of terrorist events.

Under such advised circumstances, the surprise advantage can possibly be minimized. Refinements of this type of research can well lead U.S. counter-terrorist programs beyond their current defensive posture into offensive capabilities and preventive actions where the price to the terrorists who attempt violence find the target prepared and the price intolerable.

Corporate security personnel and intelligence analysts must think like terrorists if trigger mechanisms are to be identified, understood, and explained. The process must identify both real and perceived injustices in political, social, and socioeconomic areas as visualized by the mind of the terrorists and be applicable to the corporation concerned.

In sum, through the reconstruction of scenarios of past terrorist incidents, the corporate security director and intelligence analyst becomes the terrorist in mind-set, motivation and philosophical outlook. Of course, in reality, this type of parallelism can never be truly achieved, but it makes a point.

To outwit terrorists, one must understand the terrorist mind. U.S. counter-terrorist and corporate security analysts must be exposed to those same teachings that molded the mind-set of such terrorist groups as the Party of God, Islamic Jihad, and the various factions of the PLO and Shiites.

U.S. corporate directors and intelligence analysts must now become well schooled in the mechanics for developing order of battle (Crisis Management) data and intelligence estimates and be able and willing to apply these techniques to researching terrorism.

Yet how many within your own corporate bureaucratic systems are researching Middle Eastern terrorism and/or have been exposed to or have knowledge of the Koran or other Islamic works?

Our Western attitudes are based on philosophies of Western Civilization: Plato, Sophocles, Hobbs, Locke, Rousseau, Mill, etc. As a consequence, our culture makes it difficult for our security and intelligence analysts to understand, much less predict, actions by fanatics who are obsessed with martyrdom and who have their sights set minimally on the disruption, if not the total destruction, of both global and individual corporate infrastructures.

Recommendations

To Summarize: Corporate security and intelligence departmental teams should be developing dual-track scenarios of past terrorist events perpetrated by specific groups to:

- Compare past terrorist incidents against each other and determine how such may manifest itself against each and every individual facet of a specific corporate structure.
- Compare all events with current threats by identifying similarities and indicators of early warning that will trigger Crisis Management proactively.
- Develop intelligence collection requirements beyond reporting of chronologies of incidents. Identify trigger mechanism that set off such incidents. Provide early warning to safeguard potential commercial targets.

Ongoing exposure of U.S. corporate security and counter-terrorist analysts to understanding and acquiring insight of the cultural environment of terrorists under study is crucial.

Interaction with not only appropriate government agencies, but within the corporate security world is also essential. Believe it or not, the situation that prevailed in One Liberty Plaza prior to September 11 was akin to there being in place more than a dozen security departments who seldom, if ever, met regularly, exchanged views and information, or operated any form of specific environmental collation program.

In multi-occupancy corporate buildings the absence of an all-encompassing, team-oriented, and open door exchange of information by security personnel plays directly into the hands of any individual or group seeking to exploit such an inherent weakness.

You have a saying in America that has been one the most appropriate adages I have heard quoted, something akin to: "Wake up and smell the coffee!" It is an apt recommendation used to alert the recipient to a more positive form of reaction and response when the natural tendency for paranoia, bureaucratic dogma, and negativity persists.

I would venture to suggest that an appropriate replacement statement, designed to implant the extent, import, and urgency that needs to be applied to the current situation facing corporate America would be: "Wake up and smell the carnage!"

I, along with my fellow security specialists, am truly fed up with having to state in hindsight, "I told you so!"

Security Audit Criteria
RISK & THREAT ASSESSMENTS
To include asset and personnel threat vulnerability and general corporate risk assessments, with an emphasis on counter-measure development.

PRE-EMPLOYMENT SCREENING & PERSONNEL VETTING
Thorough and intense. The days of the "fixed-fee" perfunctory inquiry are over. Negligent hiring litigative dictates will take on a greater and more punitive meaning in the days to come. "Cheapness of Application" will soon translate to "Expense beyond Compare."

PREMISES EVACUATION PROGRAMS
Evolve, test, improve, hone, and practice. Any corporate environment that does not undertake a minimum of four premises evacuation drills per annum, involving all personnel, further erodes the confidence and security awareness of its employees.

CRITICAL FUNCTION PROGRAMS
Prioritizing critical function stabilization in the event of a disaster is not an issue that can be implemented at the time of the occurrence. Detailed plans in the form of operational guides and standard operation procedure manuals should exist and be upgraded to include contact information, emergency services contacts, locations of emergency staging areas, evacuation routines, utility shut-off devices, etc.

SECURITY & PROTECTION
Critical corporate information MUST be so treated, backed-up and stored on a time-sensitive basis at all times. That will facilitate the ability of the corporation to function during the recovery or aftermath processes following a disaster. Emphasis should be on the speed of mobilization and restoration of informational and

operational assets essential for the basic functioning of a given corporate infrastructure.

TRAVEL SECURITY & SAFETY
Applicable on a priority basis to include executives at key-man level through to all employees required to travel in pursuit of the development of corporate business on a local, national and international level. Many informational assets and resources are available for research to provide time-sensitive travel advice and guidance.

EXECUTIVE PROTECTION
Key-man specialized protection for chief executives and their families is essential to incorporate both the working and domestic environments. Estate security, crisis management, emergency and medical evacuation, and (where applicable) kidnap and ransom programs need to be put in place.

SECURITY CONTRACT
Despite the current reinforcing of what has always been considered the "joke" of the security industry in general, contract security services need to be sought from those who not only enjoy an acknowledged reputation for excellence, but those who are prepared to be constantly tested and tried for quality assurance purposes. The standards and expectations of the employing corporation must always prevail. Under no circumstances should the proverbial tail wag the dog!

Personnel Awareness Training
It should not be considered an abhorrence, unethical or otherwise an abuse of liberties to gently coerce employees to become security aware of the environments in which they operate. The additional "eyes and ears" are an invaluable intelligence-gathering asset. In most instances, employees are now more receptive than ever before to protecting themselves, their colleagues, and their business environments. This is a fact that can and should be exploited to extremely good effect by competent and respected security managers.

There is nothing in the world that compares to the spirit of determined unification of purpose and intent more than that which applies to the otherwise diverse and ordinarily divisive populous of the United States of America! Memories are short, lest we forget.

Security & Safety Department Representation
Any corporation that delegates and dilutes the awesome responsibility for overall corporate security to existing alternative

departmental infrastructures—premises and facilities departments, human resource departments, etc., who are generally manned and operated by individuals unskilled in the science of security per se, and whose most pressing responsibilities and focus understandably lies elsewhere or that places a total reliance on outside vendors, thereby culpably abrogating direct responsibility, is providing a grave disservice to itself, its shareholders (where applicable), and most certainly, its work force.

Business Espionage Prevention
By John M. Elliott

How are you protecting your business?

The word "espionage" congers up many images. In today's business climate, make no mistake about it, there are countries, corporations, and persons who want to know more about your company and employees. This information is sought so as to give another company a competitive advantage over your company. How this information is obtained is defined by how you are targeted and who's doing the targeting.

A Few Definitions to Get You Started
Economic Espionage is defined as the unlawful or clandestine targeting or acquisition of sensitive financial, trade, or economic policy information, proprietary economic information, or critical technologies, excluding the collection of public domain and legally available information.

Industrial Espionage is defined as any activity conducted by a foreign government or by a foreign company with the direct assistance of a foreign government against a private U.S. company for the sole purpose of acquiring commercial secrets, excluding the collection of public domain and legally available information.

According to the annual report to Congress on foreign economic collection and industrial espionage 2001, the U.S. business community estimates that in calendar year 2000 economic espionage cost from $100-250 billion in lost sales.

Competitive Intelligence collection is a third area of concern to businesses. Simply put, this is an activity to acquire information about your business for the purpose of giving someone else's business a competitive advantage over yours, most often done legally through publicly available information, but sometimes illegally by unscrupulous collectors.

A survey report in September 2002, authored by the American Society of Industrial Security, International (ASIS), Price-WaterhouseCooper, LLC, and the U.S. Chamber of Commerce, which surveyed various Fortune 1000 corporations and small businesses during the period July 1, 2000 to June 30, 2001, concluded that losses of proprietary information and intellectual property ranged from $53 to $59 billion.

According to PriceWaterhouseCoopers' "Trendsetter Barometer" newsletter of March 27, 2002, companies placing a premium on competitor information are outperforming their peers in sustained revenue growth, gross margins, and a number of other key performance measures. Changes in pricing, new product initiatives, and changes in corporate strategy top the list of desirable information.

Yes! The spies are out there. Go to www.hotjobs.com and enter key word "competitive intelligence." I did (on 5/7/02) and 221 jobs popped up. Using the same keywords at monster.com you get 941 listings. Yea folks! The people they hire may someday soon be stalking your company.

Companies that are the most successful are those that have the best information and use that information the most effectively. Then, why give away your company information knowingly, or most often unknowingly? You've probably done a pretty good job in protecting your physical plant with video cameras, badge access, etc. What have you done to protect the intellectual capital of your company—which information is stored in the brains of each and every employee?

How Are You Collected Against and the "Antidote" (Countermeasures)

There are a myriad of techniques utilized by "collectors." And there are many things you can do in terms of countermeasures to protect your company.

It all starts with employee education—educating your employees on how they are collected against, what to watch for, what to say and do, and what not to say and do. Know where you're vulnerable. The remaining portions of this article will address some of the more common techniques used to collect against you and what to do about it.

1. Trade shows and conventions. A seemingly innocuous event, but ripe for information gathering. A skilled elicitor of information has no problem against an unsuspecting employee. This is a "target-rich" environment. The elicitor has planned his conversation. It's friendly and not intrusive. It's slick and the conversation is comfortable. During the conversation your employee has complained about a rise in their health insurance rates, or a cutback on vacation time. What does this tell the elicitor? It could provide a clue to your companies' overhead rate. It's a small part of the puzzle for the elicitor, but by the end of the conversation he has a bigger picture. Do your

employees understand what they can and cannot say about your company or product at trade shows? Have you ever debriefed your employees after they've attended a convention? It might be a good idea. You can learn whether or not they were a target of an elicitor. You can also learn what your competitor is interested in and this can benefit you.
2. Pretext conversations through Internet newsgroups, chat lines, or email. What are your employees saying about your company through these venues? Intelligence (information) collectors routinely scan these sources of information. Employees should be cautioned about what they say in these venues.
3. Technical surveillance of your home or office. Although rare and illegal, it's a technique that can yield valuable information. If you have suspicions, obtain the services of a competent "sweeper" to detect such devices. If you find a device, don't let the whole world know. You could use it to your advantage.
4. Computers. Do you have firewalls? What is the status of your access codes? Do you know where your laptop is? Hackers know all the security vulnerabilities out there. Do you? Are your computer wiring closets locked?
5. Do you have employees who write trade articles? Are these articles screened to insure they contain no proprietary information?
6. How about your SEC filings? Are you putting too much information in there? Many companies produce much more documentation than is legally required.
7. How about your press releases, brochures, website, annual report, or employment ads? Is what's in there reasonable, or does it provide a business profile that can make you vulnerable to the collector? Are you putting too much information in there, i.e., computer systems used, client businesses you're supporting, specific project names, etc?
8. Do your employees travel abroad? Have they ever received a pre-travel briefing concerning the protection of your business information? How about a post-travel debriefing? Your business information is particularly vulnerable during foreign travel periods. Foreign companies, governments, and intelligence officers don't play by the same rules.
9. Most of the time, trash is just trash. What do you do with your trash? Do employees have separate trashcans, one for personal trash, and the other for work product trash? Is it dumped in an unsecured dumpster? A rough draft of a business proposal thrown away and found by your competitor is like finding gold. Do you shred first? You can never have too many shredders. Is there an employee responsible for "trash" security?
10. Do your employees leave work product on their desks at the end of the day? Is the "in" box cleared out and secured. Easy pickings for anyone interested in what you're doing.

11. Do you conduct background checks? Conducting a background check of all prospective employees, regardless of position sought, is a step in the right direction. Advise the prospective employee of your intention to do a background check and obtain a written acknowledgment from the prospective employee. This check, at a minimum, should start with residence verifications dating back at least three years. Without this verification a criminal records check attempt is useless. How are you going to know where to do a criminal records check if you don't know where the employee has lived? Once residence information is verified, do a criminal records check of that county. Then proceed with educational verification and previous employment verification. Could the prospective employee be in fact working for your competitor? Maybe!
12. Have you ever considered having your employees execute a nondisclosure/no-compete agreement? This impresses upon the employee your seriousness about not disclosing information learned at work to outside third party persons. Further, it minimizes information about your company and its use against you. The terms of the agreement should be reasonable and constructed utilizing legal counsel.
13. Do you hire temporary employees? What do you know about them? What company information will they have access to?
14. The telephone. What a wonderful piece of equipment to be used to collect against you. Does your company receive calls from those purporting to be from a trade magazine, local media, etc? Always ask for their call back number and learn their bona-fides before answering their questions. And how about that company phone directory? It's a prized possession of any collector. I'm not advocating that phone directories be accounted for on a daily basis, but your employees should understand the importance of this information to a collector.
15. The copy machine. A great "hang-out" for someone who has penetrated your company. Are bad copies tossed into the trashcan next to the copy machine? Get rid of the trashcan. Have the employee take all trash copies to their workstation and dispose of them properly. Do you have a copy machine accounting mechanism to determine if the copy machine is receiving extraordinary use? Copy machine usage is a simple way to uncover activities that could be harmful to your company.
16. Don't routinely discuss security countermeasures with employees, especially with employees you're not familiar with.
17. Do high-level company officials always fly or drive together? Separate travel arrangements, although inconvenient, may be warranted. In the event of an accident your company will not suffer from catastrophic key personnel losses.

The foregoing information was not meant to be all encompassing. It was presented to give you a simple picture of how vulnerable you and your business can be. Just because you don't often hear about it in the news, that doesn't mean the collection of information against companies isn't happening. It certainly is, but you'd never know your company was a target until you started seeing your market share drop.

In summary, educate your employees on the importance of protecting company information. Determine what information should be protected and how to go about accomplishing that goal. Identify your vulnerabilities and close the holes.

John M. Elliott is President of the Aegis Consulting Group, and a retired FBI Counterintelligence Special Agent with over twenty-five years of federal law enforcement experience.

Taping Phone Conversations
By T. A. Brown

If an employee is suspected of selling company secrets or a business partner is suspected of being fraudulent and you have decided to tape their phone conversations as a means of gathering evidence, it is wise to know the laws on taping before you do so.

Definition of a one-party state vs. a two-party state: In a one-party state only one person involved in the telephone conversation needs to give consent to the recording. In a two-party state both parties to the conversation need to give consent to the recording. The following list of states will tell you if you live in a one-party or a two-party state.

If you live in a one-party state and are recording a conversation involving someone who lives in a two-party state, the law where the information is being used should apply. Federal law, being a one-party law, may apply when the conversation is between parties who are in different states, although it is unsettled whether a court will hold in a given case that federal law "pre-empts" state law. Remember, either state may choose to enforce its own laws. To be safe, always check with an attorney in your own state first if you are going to be taping across state lines.

If the recording is to be used as evidence in a two-party state, there must be a beep every 15 seconds to keep both parties informed that the conversation is being taped.

The research for this article was found on the following websites: Check them out for further details.
www.spyman.com/laws.htm
www.rcfp.org/taping/
www.pimall.com/nais/n.recordlaw.html

Recording conversations of employees within a business office legally varies from state-to-state and you will need to check with an attorney in your state before doing so. Some states vary by jurisdictions within the state.

Alabama: One-Party
Arizona: One-Party
California: Two-Party
Connecticut: Two-Party
District Of Columbia: One-Party
Georgia: One-Party

Alaska: One-Party
Arkansas: One-Party
Colorado: One-Party
Delaware: Two-Party
Florida: Two-Party
Hawaii :One-Party

Idaho: One-Party
Indiana: One-Party
Kansas: One-Party
Louisiana: One-Party
Massachusetts: Two-Party
Michigan: Two-Party
Mississippi: One-Party
Minnesota: One-Party
Nebraska: One-Party
New Hampshire: Two-Party
New Mexico: One-Party
North Carolina: One-Party
Oklahoma: One-Party
Ohio: One-Party
Rhode Island: One-Party
South Dakota: One-Party
Texas: One-Party
Vermont: One-Party
West Virginia: One-Party
Wisconsin: One-Party

Illinois: Two-Party
Iowa: One-Party
Kentucky: One-Party
Maine: One-Party
Maryland: Two-Party
Minnesota: One-Party
Missouri: One-Party
Montana: Two-party
Nevada: One-Party
New Jersey: One-Party
New York: One-Party
North Dakota: One-Party
Oregon: One-Party
Pennsylvania: Two-Party
South Carolina: One-Party
Tennessee: One-Party
Utah: One-Party
Virginia: One-Party
Washington: Two-Party
Wyoming: One-Party

On a similar topic, it is against the law to record sound when videotaping someone when they are under surveillance for workers compensation fraud, theft, etc. Check with your local Radio Shack or other like-minded stores for the means to disable the sound on your video camera if it is to be used for this purpose.

Business Trip Security
By Robert Siciliano

The business traveler often finds himself or herself in places domestic or international without a clue as to what the new surroundings have in store. Different cultures can bring on a whole new set of rules. Security pre-planning is the key to ensuring success.

If you are a general business traveler, these rules apply to you. However, if you are a high profile executive, in addition to the following rules, you require professional executive protection strategies, including bodyguards, bullet proof cars, and solid strategic principles.

Health Safety
Americans take for granted clean water, refrigeration and the best medical care. Be very health conscious when traveling abroad.

1. Beware of sunstroke.
2. Get your shots. Hepatitis A and B and Tetanus/diphtheria vaccinations. Malaria and other intestinal virus infections can ruin a trip or even kill you.
3. Cover your drink. Slipped Mickey Finns in the form of sedatives will surely end your trip in someone's trunk.
4. Know your options for medical care.

Emergency Planning
When you leave U.S. soil, finding help with a variety of issues such as lost passports, health problems, abductions, thefts, or any other mishaps can be an uphill battle. Language barriers and overall discontent with Americans can make getting assistance a nightmare.

Register with the U.S. Embassy or Consulate and inform them and relatives of your exact travel plans. If your destination becomes unstable, the embassy will make you aware of the current political climate.

Rental Cars
A colleague had just pulled up to the exit of a rental car lot in Spain. Within seconds a man knocked on her driver's side window and pointed to her tire and kept saying "Flat, flat, flat". She put the car in park, got out of the car, and took a look at the tire. The tire was fine, but the man was gone. She got back into the car, and her

pocketbook was not there. The thief's accomplice had opened the passenger door when she got out of the car and had taken her bag.

When she reported this to the police, they asked her if she had been a victim of the flat tire scam. The fact that the police and the rental agency knew about this scam and didn't forewarn her when she rented the vehicle is baffling.

1. Hide rental agreements since they are dead giveaways to "traveler." Keep them off the dash.
2. Don't store valuables in the trunk. Many rental cars use the same keys.
3. Don't pull over for anyone for any reason. Bumps and distress signals are a common ruse.
4. Taxicabs are not always what they seem to be, so exercise caution.

Securing Your Home While Away

Whether you are planning to be gone for only a few days or for months at a time, you must secure your home before you leave.

Here are four important points to remember.

1. Tell only those who absolutely need to know that you'll be away.
2. Arrange for a neighbor or trusted friend to keep an eye on your home, and gather mail, newspapers, and the occasional flyer stuck in your door or your mailbox. Don't stop deliveries of mail and newspapers because your name and address ends up on a "stop list," which can become a resource for burglars.
3. If you have an answering machine, don't change your outgoing message to "Gone to Vegas on business. See you in two weeks." Maintain a professional message stating your name and telling the callers to leave a message at the tone.
4. Don't tempt thieves by visibly packing your car the night before you leave.

Safety On The Streets

Over the years I've noticed that all victims have one thing in common: they aren't aware of what's going on around them. Security experts agree that the best way to protect yourself is to avoid potentially dangerous situations. But even if you must be in a dangerous situation, a well-rehearsed plan of action can prevent you or your family from becoming victims.

Understand the fundamentals—body language, awareness, and intuition. Pay attention to the way you walk, your posture, facial expressions, and eye contact. Know what is going on 50-100 feet around the perimeter of your body at all times. Pay attention to the feelings inside and around you. When the hair on the back of your neck stands up, watch out.

Here are some points to remember when you walk alone.

1. Be alert and aware. Continually scan the area around you. Walk briskly, keep your eyes and chin up, and listen for warning sounds.
2. Choose a sensible and safe path to or from your car. Use well-lit, populated routes whenever possible.
3. Be ready to unlock your car without having to fumble for keys.
4. Avoid poorly lit areas and shortcuts where you are hidden from others.
5. Don't appear to be lost or alone. Assailants often prey on people who seem to be unfamiliar with their surroundings.
6. Be ready to react. Keep your hands free and wear sensible clothes and shoes that do not restrict your movement.
7. If you think someone is following you, change direction, cross the street, slow down, speed up, or go into stores.
8. If the person keeps following you, immediately ask for help. Ask a security guard to escort you to your car.
9. Don't assume that you are invincible or immune to assault. Never feel embarrassed or awkward about asking for help.
10. If you use a weapon, carry it in your hand and be prepared to use it. Or look around for rocks, bottles, sticks, or clubs that you can use as weapons if necessary. Carry Mace® or pepper spray if your state or city allows it.
11. Carry 10 or 15 one-dollar bills in an easily accessible pocket. Then if someone tries to rob you, you can throw the "chump change" several feet away.
12. Keep your ID and other items containing your address in your pocket, not in a purse or wallet.
13. Carry only the cash, checks, or credit card you absolutely need.
14. Do not flaunt money or valuables. Do not wear excessive jewelry. Turn rings around to hide gems and avoid wearing loose gold chains.
15. Don't use an outdoor automated ATM machine at night or in unfamiliar or unsafe surroundings.
16. Be alert when leaving stores or shopping malls. Shopping time is prime robbery time.
17. Stay alert! If someone moves inside your personal comfort zone, move away. If he persists, run. If necessary, strike him before he strikes you.

18. Do not use the stairwells in parking garages. Try walking down the auto ramp instead. As long as you watch for cars, the ramp is much safer.
19. Walk facing oncoming traffic. Walking against rather than with traffic reduces the risk that you will be followed, forced into a car, and abducted.
20. If someone in a car asks for directions, stay far enough away from the car so you can turn and run easily. Or simply say, "I don't know" and keep walking.
21. Use extreme caution when you are approaching any unlighted entryway. A common tactic of criminals is to remove, unscrew, or break the light bulbs in such places.

Learn The Culture

Looking and acting like a tourist is an excellent way to attract thieves. On the other hand, if you blend in with the local population, thieves won't notice you. As the saying goes, "When in Rome, do as the Romans" Dress appropriate to the culture. Flashy Americans walk around with a bull's-eye stamped on their head. Adapt to the culture of style and dress. Do everything possible to blend in.

1. Familiarize yourself with strange cities, countries, and cultures before you leave home. At your local library, bookstores, and travel agencies are dozens of books, pamphlets, travel guides, maps, language guides, and other information on any area of the world.
2. Memorize key locations in the area you're visiting. And write down the addresses and telephone numbers of places to contact for help if you need it—embassies, police stations, hospitals, airports, train stations, and so on.
3. Ask your hotel concierge or the local police which areas of the city should be avoided, particularly after dark.
4. Dress like the local people. Don't advertise the fact that you're a tourist by dressing in expensive vacation gear if everyone else is wearing jeans and T-shirts.
5. Check local laws about carrying Mace® or other defense weapons. Be aware that these cannot be carried onto a plane, although some airlines allow you to include them in check-on baggage.
6. Contact the U.S. State Department to learn about the political climate of your travel destination. Are the citizens hostile to Americans? Are there ongoing political or economic tensions that could erupt into overt hostility while you're in the country?
7. Don't offend the locals because of ignorance of their customs. You will not only be thought ill mannered, but thieves will also

spot you in a minute. Before you arrive, learn enough of the language to communicate essential ideas. And learn the basic manners and customs of the area so you don't run the risk of being impolite.
8. Be able to recognize the uniforms of the police in the cities you will visit.
9. Be wary of offers of friendship or excessive attention from strangers, guides, and others you meet.

General Travel Tips
1. Carry traveler's checks and credit cards, not cash. If you must have cash, carry as little as possible.
2. Place valuables, one extra credit card, emergency cash, and emergency telephone numbers in a hotel safe.
3. Don't make a display of material wealth or be a heavy tipper.
4. Make two photocopies of the first two pages of your passport. Put one set in the hotel safe and one set in your luggage.
5. Don't wear clothing or items that obviously identify you as an American.
6. Don't wear or display items that identify your company. These are marks for terrorists and kidnappers.
7. Bring bottled water with you when you travel abroad.
8. Avoid overindulging in alcohol or anything else that impairs your judgment.
9. During meetings keep all belongings on your person especially when you travel room to room or to the bathroom. Professional thieves scam these areas a day into the meetings once the attendees are comfortable and feel safe in their surroundings.

Safety In Hotels
While the streets of a strange city may be dangerous, hotels can be even more so, perhaps because they feel like places of relative safety to a tired traveler.

Choose a reputable hotel, preferably American, located in an area where crime is low. Reserve a room on the second floor or higher, preferably in the middle of the corridor, and not too far from the elevator or staircase.

1. Take a room that faces the street or that overlooks a swimming pool or other activities area.
2. When you are reserving a room, if the desk clerk blurts out your room number so others can hear, quietly request a new room.
3. Have a bellhop take your bags to the room. In this way you can ask the bellhop to inspect the room before you enter. Check

under the beds, in the closets, in the shower, behind curtains, and anywhere else someone might be hiding. Check immediately to be sure that all the locks are working properly.
4. If there is no bellhop, ask the manager to accompany you to your room.
5. Whenever you enter your room, secure the dead bolt and chain locks. Make sure the peep hole works, and before unlocking your door use it to verify the identity of maids, room service attendants, or anyone else who knocks.
6. Portable travel locks, motion alarms, door braces, door jammers, and rubber wedges are available. Buy them and use them.
7. Keep windows and balcony doors locked.
8. Be aware of suspicious people in the elevators and hallways.
9. Leave nothing of any value in your hotel room while you are gone. If you think for one second that your laptop computer, the information on your laptop, your jewelry, money, or anything else that has significant value to you is safe unattended in a hotel room, you might be delusional.

Think, too, how easy it is for a man in a three-piece suit to walk into your room while it is being cleaned, to say to the maid, "Excuse me, I just have to get something," and then grab a suitcase with all your camera equipment.

1. When you leave the hotel for any reason, give your key to the front desk attendant. Doing so eliminates any possibility of losing your key, which may be stamped with the name of the hotel and your room number. Additionally, don't put your room key down beside you at a restaurant, bar, or poolside. It can be easily stolen.
2. In case of fire, pre-plan an escape route by memorizing the location of your room in relation to stairways. Don't use the elevators. They can malfunction in heat and trap you.
3. Keep valuables and warm clothing at your bedside in the event of a hotel fire. Better yet, keep valuables in a hotel or room safe. Hotel fires occur with greater frequency than many people realize and in other countries the building codes and fire regulations may be far less stringent than in the United States.
4. Be suspicious of a call from the front desk just after checking in requesting verification of your credit card number "because the imprint was unreadable." A thief may have watched you enter the motel room and called from the guest phone in the lobby.
5. Look carefully into elevators before entering. Stand next to the control panel to access emergency alarms or to press all the buttons so you will get off at the next floor.

6. Don't leave your door ajar while getting ice. Carry your key with you.
7. If someone is following you down the hall, let him or her pass you before you open your room door.

Safety In Airports

Airports are another haven for criminals. In the parking lot lurks the car thief and the mugger. Throughout the terminal are the scam artists and the pickpockets. In the baggage claim areas are the baggage thieves.

1. If you must check your luggage, wait to see it go into the "chute" after it is taken from you. When booking your reservation, get a seat at the front of the plane so you can get to baggage claim quickly.
2. Store your carry-on luggage across the aisle instead of over your head. You want to keep an eye on it. Otherwise someone can easily go into the overhead bin and remove your belongings. Never put a pocketbook under the seat. The person behind you can remove a credit card and you might not know it for a couple of days.
3. When riding in a shuttle bus, don't let your luggage leave your side no matter what. The rear compartment can be opened while the bus is stopped at a traffic light. People are constantly getting on and off the bus at different stops just to steal luggage.
4. If you are a salesperson or a high-profile person, such as an executive, celebrity, or politician, you might consider putting a fictitious name on your luggage so thieves don't target it.
5. Because tags fall off or get ripped off, put photocopies of your passport, I.D., and itinerary in your luggage. In case a bag gets lost, someone who recovers it will be able to forward it to you.
6. Place any baggage, laptops, or briefcases on the counter in front of you when you stand at rental car, hotel, and airport ticket counters. If you put these at your feet to the left, right, or behind you, you become a prime target for distraction thieves. For example, a very emotional person walks up to you while you are waiting for the clerk at the counter, asks you how to get to the Alamo, and then starts to cry. In the confusion an accomplice sneaks up behind you and removes the laptop that you placed on the floor next to you.
7. Don't take your eyes off your belongings while they are going through security or screening checkpoints. This is a prime location for distraction thieves to steal laptops, pocketbooks, and briefcases. Once you put your belongings on the movable belt, one thief distracts you from immediately going through the metal detector by either dropping a handful of change, causing

a scene, clipping a metal object to the back of your coat that will cause a delay, or saying, "Hey, don't I know you?" Anything to keep you from going through the metal detector for 30 seconds while the accomplice walks through clean and picks up your belongings. If you become distracted for an instant your valuables are gone!
8. Never leave your bags unattended. They can be stolen or rarely, but worse, someone looking for the opportunity could hide bombs or drugs in your bags.
9. Do not overstuff your luggage. It can pop open easily. In addition, stuffed luggage looks to a thief as if there might be something of value in it.
10. Don't use fancy, expensive luggage.
11. Put all electronics, cash, jewelry, medicine, and important papers in your carry-on luggage.
12. Ignore other people arguing and strangers who are overly friendly. These could be staged distractions to make it easier for a thief or pickpocket to rob you.
13. Be alert to anyone in the baggage claim area paying undue attention to you.
14. Don't let anyone help you with your airport locker. Someone might insert quarters for you to appear helpful but then give you a different key without your knowledge.
15. Be aware of any contact with others, even if it is a good deed they are doing. They could be setting you up.
16. Request window seats in a plane's coach section. Hijackers often take hostages from first-class aisle seats.
17. If your plane is hijacked, do not make eye contact with the hijackers, which can increase the chances that you will be singled out for attention. Calmly rally other passengers and use strength in numbers to pounce on your attackers.
18. Don't tell a stranger your plans. The accomplices of hijackers often disguise themselves as passengers.
19. In foreign countries arrive at the airport as close to departure time as possible and wait in an area away from crowds. If a bomb has been planted, it is more likely to go off in high-traffic areas.
20. When renting a car in a foreign country, rent one that is common in that country. Make sure that any rental car in America or abroad doesn't have "rental" written all over it.
21. Guard your passport and tickets carefully. There are many undesirables looking for a new identity who would love to have your passport.

Robert Siciliano is a Boston based Professional Speaker, Personal Security Consultant, and president of three security related companies. He has 18 years of security training in martial arts, personal body guarding, barroom bouncing and observing the human condition. He is the author of two books including *The SafetyMinute: How to be safe on the streets, at home and abroad so you can save your life!* His seminar topics include; Safe Travel Security, ID Theft Security, Workplace Violence, Nurse Security, Realty Security, Self Defense, Children Security and Public School Security. Robert has appeared in Mademoiselle, Good Housekeeping, Consumer Digest, the NY Post, and Boston Herald, and he has been on national TV on CNN, MSNBC, FOX, the Montel Williams, Sally Jessy Raphael, Howard Stern, David Brenner, and the Maury Povich talk shows.

Disaster Management PS11
By Kevin J. Ripa

September 11, 2001 started like any other day, but it ended quite unlike any other day in recorded history. Disaster Management (DM) for most companies was a liability before September 11. It was the same as having a bodyguard for a corporate executive. It was a corporate cost that had no quantifiable way to pay for itself. Disaster Management PS11 (Post September 11) is viewed with an entirely new respect. The tragic loss of life was horrendous and unparalleled. At the same time there was another far-reaching effect of the terrorists and that was the thousands of companies that lay in tattered ruins. Single office companies simply ceased to exist, while global corporations with head offices in the Towers became like octopi without heads.

Imagine the corporations that relied on the offices in the Towers to manage their operations. In some cases server rooms housed the entire computer backup for these corporations. All computer data was lost. All paper data was lost. Backups were lost. Client information and records were lost. In the case of highly sensitive computer data that couldn't be backed up, such as highly classified CIA data, their hard drives were recovered from the tangled wreckage where possible, and forensics experts were called in to restore the data from the drives that in some cases had taken months and even years to compile. In some extreme cases, experts were watched over by agents, and when they found something of importance, they were not allowed to monitor the transfer of the file to the new medium.

Disaster Management (DM) is now no longer considered a money drain. It is considered a necessary cost of doing business in today's world. It encompasses so much of the act of business that it would be irresponsible to treat it as an aside to business. To not make DM an integral part of your everyday business on the corporate level, whether big or small, is to show that you have learned nothing about the misfortunes of others. Odds are very good that you are safe from a terrorist attack, but quite technically the components that went into the destruction are commonplace. As we look at the components, don't just think of them in the context of a terrorist attack. Any of them could occur individually with catastrophic results to a business. Let us look at these components.

Plane crash: Computer data on corporate laptop of a passenger that has not been backed up. The passenger could be an upper level executive that didn't feel it necessary to share certain integral machinations of the company with anyone. The other side of the coin would be an aircraft crashing into your office building. It certainly would not have to be a large aircraft to destroy the building that houses my office.

Explosion: This happens daily around the world.
Fire: This also happens daily around the world.
Water: From sprinkler systems and burst pipes.
Surges: Electrical surges destroying computer equipment.
Collapse: Not as commonplace as fire and explosion, but certainly a possibility.
Loss of Life: There is no question that every life is important, but what about the loss of someone who was the only one with an integral piece of data or instructions?
Downtime: Anytime disaster strikes, an important part of management must include the loss of productive time.

There can be no mistake that each and every one of the above components could potentially bring a company to its knees. There are other components as well that individually could play a role in the downfall of a business.

Flood: Instances of flooding are commonplace.
Temperature: On a Friday night your building has a malfunction in the HVAC systems that regulate the temperature in your office/building. In Phoenix this could be disastrous because the temperature spike would cause air-conditioned server lockers to overheat, damaging computers. In Detroit in the winter it could cause pipes to freeze and burst. Speaking of freezing, have you ever seen what happens to an LCD computer screen when it freezes? LCD stands for Liquid Crystal Display. The key word being Liquid. When liquid freezes it expands. Have you ever broken a calculator display and seen the black leak across the screen? Same thing basically happens when the cells freeze. They expand beyond their membranes and burst.
Quarantine: From Legionnaire's disease in a hospital or high occupancy area to an anthrax scare, whether real or a hoax, it must be dealt with.
Cyber Attack: A competitor or a 12 year old (maybe hired by the competitor!) decides to launch a Denial of Service attack against your network. Worse yet, a malicious entity hacks into your network and deletes the contents of a storage server.

Electricity: We take this for granted these days. What happens to computers that suffer electrical loss? Without a surge protector, you can potentially corrupt the hard drive. It isn't very good for them when the spike from the return of the power occurs either. Extended outage can also contribute to the temperature example above.

I certainly hope no one is out there thinking that any of the above can't happen to them. Having said that, I am sure there are those of you who think DM is for the big guys, that it is only necessary for the large corporations to worry about that stuff. If you consider the above risks in the proper context, you will see that any of these could apply even to a one-person operation in the basement of your home. Even the loss of life could affect a one-person business. Imagine you run your office out of your home. You have a number of ongoing contracts and jobs in progress. You drop stone dead one day. Where would your spouse start to sort out your business affairs if you had no DM plans in place?

Now that I have your attention, we will look at some of the things that you can do to mitigate the normal recovery of business operations after disaster. Obviously we cannot plan or discuss every eventuality in this forum. What we will do instead is lay out some guidelines that you can follow from the perspective of the single person office on up through multi-location corporations. You will see the recommendations from both ends of the spectrum. It will be up to you and your management team to discuss what you feel is appropriate to your circumstances.

The number one thing you have to contend with in today's wired world is the fact that virtually everything you do is on computer. If it weren't, you wouldn't be reading this book. It would stand to reason that some of your most vital data would be contained within your computer hard drive. In a great deal of instances a company would be inconvenienced by a catastrophe as long as their computer survived intact. Conversely, the same company could be completely devastated by simply having a destroyed computer. If I seem to be harping on the computer end of things, it is because I am!

Protecting your computer media should always be foremost in your mind. Whether it is a single computer or a thousand-computer network, the rules are the same. Backup, backup, backup. There is nothing more important than having backups of your hard drive to fall back on. This applies to everything from a simple hard drive failure to having your entire computer destroyed by fire. Something to consider when exploring backup options is to plan where you will keep your backups.

In a corporate environment, whether Small Office/Home Office (SOHO) or large corporation, you should have two sets of backups. One backup would remain resident within your work environment. This should be backed up at the end of every day. There are a number of different methods of doing this. The other backup should be an offsite backup. In the case of a SOHO, it could be a removable hard drive that you backup once a week. The key is to keep it physically away from the structure that your office is in. If the building housing your office were to be completely destroyed, as in the WTC case, you could, at the very least, have your client list off the backup by the next day, informing them that there may be some delays.

In the case of multi-terminal offices and/or corporations, there simply is no question that you should be employing offsite storage facilities. (This is also available to SOHO). It is extremely convenient and allows for backups at a predetermined frequency that is completely unsupervised by you. Some of the largest offices in the WTC were worldwide headquarters that housed the computer network operations of their offices around the globe. During the initial impact, the aircraft destroyed a number of server banks, effectively throwing global offices offline and into the dark. Thanks to the forethought of offsite data management, the prepared corporations had their entire computer networks (less WTC of course) back up and running before the first building collapsed.

Not to continue harping on the computer end of things, but another consideration we allude to be electricity surges. What happens if the electrical sub station that supplies you has an explosion? Initially, it could send a surge up the line that would destroy any sensitive computer or electronic equipment. A relatively cheap insurance policy would be to install surge protecttors and uninterruptible power supplies.

Redundancies are the order of the day. We spoke of backups already. If your computer information or network, such as an online business, is mission critical, you may need to consider something like mirrors to your system (basically another computer that looks exactly like yours, if yours fails), and redundant back ups. According to Q9 Networks, a Managed Internet Infrastructure company located in Calgary and Toronto, there are nine steps to consider when determining if you need services such as this. They are:

1. 100% Internet availability
2. 100% power availability
3. High security
4. Advanced fire detection & suppression
5. Stringent climate controls
6. Customer control and visibility
7. Scalable infrastructure
8. 7x24 monitoring and maintenance
9. Access to Internet expertise

We will now move on to other DM considerations. Whether you are SOHO or a large corporation, or anywhere in between, you need to give some thought to what you would do in the event of a fire or other disaster that removed the physical site that you work in. If you are a corporation that has an entire floor in an office tower, you should definitely consider where everyone will work.

From the safety perspective in a disaster, any sized business needs to have a plan in place to collect, locate, and otherwise account for their staff. You have to be prepared for an inundation of telephone calls from concerned relatives wanting immediate information. This can take the form of an emergency cellular number that staff have and is only to be used in case of emergencies. You can also have a meeting place across the street or across town. Even an email address for the purpose can be used for displaced staff members to report their status in disasters of the magnitude of WTC.

We have all heard the stories about the person down the street whose spouse died suddenly. If this happens to you, even in a one-person operation, what plans are in place to contact clients of ongoing projects? What plans are in place to deal with the issues in progress and continue or shut the business down? This can affect any size of business if you place anyone in a position of complete authority over something. If a computer holds sensitive data that only one person has the password to, this could create a problem if that person dies. We spoke of redundancies in computers. Workplace tasks are the same.

The sad fact in today's day and age is that people don't care about your misfortunes. If they can't get it from you, they will get it elsewhere. Tragically, if the "elsewhere" serves them properly, they will not come back. Putting a strategy together to deal with disasters, big and small, need not be rocket science. Sit down and think about how each part of your business unit would be affected by the various things that could befall it and design a reaction plan to minimize the effect these things could have. In the famous words of Louis Pasteur, chance favors the prepared mind! Plan for the worst, hope for the best, and the rest is just gravy....

Chapter Two

Employee Screening

Employee Screening Through Handwriting Analysis

By Treyce Benavidez

Currently there are over 8,000 major U.S. corporations that use a hand-writing analyst for employment purposes. Let's say that you are the head of Human Resources for a Fortune 500 company. The work productivity depends upon teamwork and honest, hard working employees. You currently have an opening that needs to be filled ASAP and your Sunday ad just brought in 200 applications. Overwhelmed you begin sorting through them, spending countless hours and probably a few late nights just to find the ones that qualify for an interview. Meanwhile, your workload continues to build and you feel stressed and frustrated because of it. You find 50 that will be interviewed by the department manager. Now she has to abandon her workload in order to schedule interviews and each candidate is allocated one hour. That's a total of 50 hours—more than a normal workweek. She is now stressed and frustrated, and eventually all the candidates seem to blend. Chances are that most of the candidates weren't worth the time spent interviewing them. Best solution to avoid this scenario? Call a handwriting analyst.

Companies spend approximately $3,000 on each new employee. The man-hours and money spent are counterproductive to the company and unfair to the other employees. Avoid this from occurring initially by hiring a handwriting analyst. Within hours (not days of endless interviews) you will know who is dishonest, responsible, violent, a hard worker, drug user, and more as it pertains to the job. Say "no" to added stress and workload and say "yes" to Strategic Options for Requirements and Training (SORT).

A SORT service can do all the screening of applications and provide you with only the candidates that fit the job description. After all, anyone can look good on paper and interview well. With SORT nothing is read, but measured instead are the 300+ strokes, slants, baselines, and more to see the "real" person behind the pen along with their work ethic. Many European countries require an analyst screening before hiring a maid, landscaper, or police officer.

There are more uses for employers. For example, two people qualify for the same promotion. Which one will do best? You may have to lay off one or more people. Does one have the potential to come back and retaliate? Be prepared beforehand by asking an analyst.

I just finished a case for a large automobile manufacturing company. One employee was writing anonymous notes to another employee. The "victim" perceived them as threats, but there were mitigating circumstances. The company quickly became annoyed since the notes were clearly a nuisance and not a crime. Nevertheless, they wanted the notes to stop, so they hired me. They faxed me the original applications of all employees who had access to the "victim's" locker along with the notes. There were eight employees in question. Upon receiving the fax and quickly glancing I could immediately see who the guilty party was.

However, I always do my job objectively. First, I analyzed the note and then turned it over. I repeated this same process with each application. Upon the final profile, I compared them to the note's profile and one was identical. At that point I knew I was right. As it turned out, the author of the notes had been an excellent employee for ten years. Why did he write them? His profile revealed several things, such as (1) he's "feminine" and had been teased as a child—perhaps the victim teased him, (2) he tries very hard to be a good person, which explains why he's an excellent employee, (3) his childishness shows with writing notes and putting them in a locker. This is high school behavior. Now the company knows who is involved and can deal with it accordingly. This works equally well with stalkers who are oftentimes acquaintances of the victim so that, based on the personality traits, the victim can almost always give the stalker a name.

Another recent case involved a woman who wrote a check for $500 at a casino and then told the bank it was a forgery. I was called and, after completing my analysis, I could see the dishonesty, deceit, and her discomfort at writing it. When she was confronted she stated, "I didn't want my husband to know I was at the casino because I was just released from in-house treatment for gambling addiction." Just imagine the comfort, to a potential bank customer, to know that one of the services offered is access to a handwriting analyst in case of credit card or check forgeries. This works the same way for forged wills and the like.

In all cases consistencies always reveal the true story. In fact, consistencies in an individual's writing are formed at age 13 and stay for life. An untrained eye cannot see them, but an analyst can. Consistencies will always tell the truth.

Ever had a problem with employee theft? Was someone's wallet stolen? A teenage boy at a group home said his was. Staff searched the building and found nothing. They interrogated every other boy and received no answers. Not knowing what else to do, they turned it over to the local police and put everyone on house arrest. Two weeks later the police hadn't even acted on it so they called me. I told them to have everyone, including the "victim" and staff write the same sentence ("I did not steal (John's) wallet") and sign their name. After receiving the 16 pages I went to work. Immediately I noticed that the victim himself did it. Nobody stole his wallet. I asked myself "why." I discovered in his profile that some of his traits included deceit, lying to avoid problems, and hoarding money. I called the lead counselor and asked her if, by chance, his restitution was due the day his wallet disappeared. She confirmed that it was. I then told her that "John" stole his own wallet so he could keep the money. Sure enough, he confessed. Later the wallet was found planted along the house. In a matter of hours the case was solved.

Insurance fraud is a big problem in this country. An analyst can be used to review the written incident statements and look for signs of deceit. These areas can then be investigated further. For health insurance an analyst can check for pre-existing health conditions.

Landlords can screen prospective tenants. Doctors can screen annoying patients to see if they are hypochondriacs. I receive many calls from psychologists who have been working with a client once a week for two years and they seem stuck in limbo. Their client will only reveal what she/he wants the psychologist to know. The psychologists can then fax in their writing so I can tell them what the real problem is and where to focus the treatment. All these things are done without revealing any identifying information—everyone stays anonymous. The analyst does not need, nor even care, what an individual's name is.

Defense lawyers hire an analyst to pick jurors that will be compassionate for their client, and vice versa. Teachers can screen the bullies, the kids in alternative programs, and the withdrawn children. Parents can find out what's going on with their child and the type of friends they have. Police departments and investigators can have an analyst look at suspect and victim statements to find lies. Whatever the need, a handwriting analyst can provide very valuable insight.

Handwriting Analysis is applied psychology and cannot be "professionally" learned only by a book. It takes years to become highly skilled, and the uses are endless. There are some things handwriting analysis cannot reveal, and they are age, gender, race, appearance, handedness, religion, and it most certainly cannot predict the future! It does, however, predict with astounding accuracy how an individual is likely to react to a given situation. A lack of knowledge can cause people to classify handwriting analysis next to Palm Reading and Tarot Cards, but I can only say one thing. If you want to know your future, call Ms. Cleo, but if you want the most valuable tool for profiling, call a handwriting analyst. In fact, I can guarantee that there is at least one way for handwriting analysis to benefit every person and every company.

Treyce Benavidez of The Handwriting Company received her certification for Handwriting Analysis in 1985 and her Handwriting Formation Therapist Certification in 1986. She was an agent for the former U.S. Fugitive Service and has over 10,000 field investigation hours. Ms. Benavidez is a qualified expert witness in a court of law and an international trainer for law enforcement, human resources, educational, and judicial agencies.

Background Checks in Canada
By Kevin J. Ripa

In order to conduct a background check in Canada, non-Canadians must dispense with their traditional idea of background checks. Canada operates differently than the United States with respect to information, and we do not have the extensive databases that are so readily available south of the border.

In Canada we have ten provinces and three territories. These can be compared to states. From West to East, provinces first, then territories, they are as follows (with capital cities): British Columbia (Victoria), Alberta (Edmonton), Saskatchewan (Regina), Manitoba (Winnipeg), Ontario (Toronto), Quebec (Quebec City), New Brunswick (Fredericton), Prince Edward Island (Charlottetown), Nova Scotia (Halifax), Newfoundland (St. John's), Yukon Territories (Whitehorse), Northwest Territories (Yellowknife), and Nunavut (Iqualuit). Our national capital is Ottawa, which is in Ontario. The territories span the entire width of Canada and are directly north of our provinces. The Northwest Territories extends to the North Pole. Enough with the geography lesson. So what is available?

In Canada we are very much restricted to provincial boundaries when it comes to searches. Each province differs enormously with regard to the information it allows access to. Alberta is one of the most liberal provinces when it comes to information, while Quebec is all but draconian. This article is written with Alberta rules in mind. Rules change across Canada, so be sure to check with local investigators to determine exactly what is available.

The first thing that requires mention is our Social Insurance Number. This is comparable to a U.S. Social Security Number. It is nine digits in three groups of three, and generally speaking, you can tell by the first digit, roughly where it was issued. For example, if the number begins with a one, it was usually issued in Newfoundland. You can use that rule of thumb across Canada in that the first digit gets larger as you move farther West across the country. British Columbia numbers usually start with 6 or 7. A 4 usually (but not always) denotes a landed immigrant.

Our Social Insurance Number it is extremely difficult to obtain by any means! This number is issued for tax purposes only and is not required to be given for ANYTHING else. Hence, it is not available through searchable means.

Motor vehicle information that is available includes the ability to search for a demographic (driver's license info) with just a name. You can also search for all vehicles registered to an individual or company, as well as a history of ANY vehicles EVER owned by the above. You can search for registered owner of a vehicle with the License Plate or the VIN. Driver's abstracts are also available. This applies to Alberta and Northwest Territories, as well as to Manitoba and Ontario. British Columbia has very strict rules regulating release of motor vehicle information, making the gathering of that information extremely difficult. Saskatchewan and Quebec simply do not allow these searches at all. I cannot comment on any provinces not mentioned, as I don't know. Motor vehicle information is on a province-by-province search. There is NO national database such as the U.S. DBT. By the way, a general rule of thumb applies when I state that you cannot gain access to certain information. That rule of thumb is that this article is written from what is an officially available standpoint. The reader can interpret what they will. As with anything, call a local resource for the last word on what is or is not available.

Financial searches are available for appropriate reasons and require a signature from the subject, as in the United States. As in the United States, they are available otherwise, but the requesting agency faces the peril if caught. There are two credit-reporting agencies in Canada—Trans Union and Equifax. Trans Union is the easier and less stringent of the two for access to information. However, you get less information generally, and it has, in my personal experience, been unfortunately quite rife with inaccuracy. These searches are Canada-wide and require a name, date of birth, or Social Insurance Number, and at least some address where the subject may have lived at any time in the past. In certain circumstances, this search can be done with just a name and address, but this is not common. As a point of information, we have no such thing as a credit header. You get the whole credit report or nothing at all. Having said that, our credit reports show addresses going back forever (it seems) as well as employment history.

Personal property searches are available province by province. These include anything (outside of real estate) that has a lien against it, for example, a car loan, or furniture loan, or maintenance enforcement's, or judgments. This is searchable by name. I do not know what accessibility is outside of Alberta, but I believe it to be quite easy to obtain.

Bank account information across Canada is the same as in the United States. Stay away from it without signed permission.

Corporate searches can be done against a trade name or numbered company name. This can be done for sole proprietorships as well. It is a province-by-province search and is accessible across the country. You need the company name for this search. There is NO WAY that I know of to search against a person's name to see what companies they may be involved in.

Real estate searches are available on a province-by-province basis and access differs from one to another. This is only searchable by address or property description (in rural cases). It can be searched by name within a known taxable municipality (like a county) by knowing the appropriate contacts within that jurisdiction. There is no broad search that I know of that can provide property listing by name of owner.

Court searches include the following: Civil (defendant or plaintiff), Criminal, Divorce, and Bankruptcy. Unfortunately in a number of provinces, these are not province-wide searches. They are jurisdictional. For example, Alberta alone has eleven jurisdictions. Each jurisdiction has different rules, and they vary across the board. For example, Saskatchewan is searchable province-wide from one terminal, but you need the subject's consent. Manitoba is searchable province-wide in all categories at once for a very nominal fee. In some jurisdictions nothing is searchable. Again, consult local resources. In some cases the Criminal search is only possible with an existing File Action number or a signed consent.

This pretty much sums up searches in Canada. We have a very limited number of Information Brokers within Canada, and I have only used one in Ontario. I know of another in British Columbia. I cannot vouch for their effectiveness, nor the way they do business, so I will not make reference to them here.

Obviously we all have our lists of contacts so you can read into that what you will. This article includes only those searches that any agency has reasonable and normal access to.

Should you need a referral within Canada, feel free to contact me via any of the means under my listing on the contact page.

International Resources
By M. Ettisch-Enchelmaier

This document outlines the sources available online and offline around the world needed by inhouse investigators for helping business owners to do background traces for pre-employment screening and producing credit reports on companies or individuals. These sources are discussed in the context of an international area showing to what extent these sources are available in different countries.

What sources, particularly online, are used to undertake various investigations in different countries around the world

The following essay is based mostly on information provided by investigators around the world.

In many countries telephone directories are available both as books, and CD-ROMs, or online, e.g. in Germany, U.S., and Great Britain, which is often only the Yellow Pages (classified directory) and not the White Pages general telephone directory: www.phonenumbers.net/usa.htm, www.teldir.com/eng/

In the U.S. one may search for companies and private individuals by inserting the telephone number at hand (reverse phone numbers searches) online, and also find their email address: www.freeality.com/findet.htm.

In Australia there are reverse phone discs that come out every six months.

In Germany and in Great Britain such searches, even on CD-ROMs, are illegal because of the Data Protection Act regulations prevailing in these countries.

In Mainland China there is nothing yet online except for telephone directories, and these are in Chinese. The investigator has to handle the investigation mostly by leg work.

In Switzerland telephone directories are public, although until recently a user had to be a subscriber of the Swiss Telecom to get telephone numbers both online and offline. For foreign users their national telecom company had a special reciprocal agreement for the access to this data.

The Swiss commercial registers are public listings of the registered companies as well as the names on the Boards of Administration and Board of Management, also Bankruptcy. These are not centrally available nor are land registers, which are public, too. The researcher must go to the appropriate register, but at times a written application suffices.

In Germany land (or real estate) registers are not publicly available except when a justified interest is supplied, such as buying, granting a loan, or information about assets as needed by a creditor.

In Iceland a database (Landstraust Hf, Thveholt 14, 105 Reykjavik, Tel. +354-550 9600, formerly part of the Icelandic Chamber of Commerce) has access to a company's registry and land registry, both online for a fee, as well as a "national" registry to search for names and addresses of individuals or companies. This search is online and free.

The land registry shows who the owner is, what kind of property it is, and its size and registered (not market) value.

In some parts of the USA, such as in New Mexico, property records are not yet online. In Greece databases supply information on companies online for a fee. e.g. Icap SA and Alpha MI, both in Athens.

The same applies to Germany, at times for private individuals too, e.g. Bürgel/Aachen, Creditreform/Neuss, Dun and Bradstreet/Frankfurt. Telephone directories are also available online.

There is very limited information in the Middle East or Africa and in Mexico only telephone directories are online. This is in contrast to the U.S. which is about three years ahead of Germany and 1-2 years ahead of Great Britain.

There is a large amount of access to public records in the U.S., some of which are online. Property records are public, as are court records (unless they involve juveniles), and even traffic court records are public. Court records are at the level of the magistrate (city court), district court (county or state), and federal courts including the supreme state and federal courts.

Income tax records are not public, but they may be subpoenaed from individuals. Property tax records are public in some states. Adoption records are public in some state but not in others.

A criminal case at the police department is something a researcher may get a copy of if it involves city police and state police. This does not apply to a lot of federal investigations until after they have gone to trial. After the trial the information is usually available. Once it has been testified to or used as evidence in a court proceeding, it becomes public. Some driving record information is publicly available, and in New Mexico the driver's history of accidents is online, but it does not include pictures from the licenses.

Checking credit worthiness of a potential client (company or individual)

The U.S. crimetime.com has, for this sort of investigation, a free bankruptcy check. Corporations may be investigated by state. Corporations and UCC are usually on the secretary of state home page for each state anyway, but crimetime is easy to use as the start page.

www.businesscreditusa.com used to charge $5.00 for a hit, but it is now free and will also rate the company's creditability between a-f. In addition, they also supply the name of the owner(s).

In Australia databases supply a lot of information on companies to their members for a fee.

In Great Britain there are three leading databases (Equifax, Experian, Dun and Bradstreet) supplying information on companies online for a fee. But there is also the Companies House (Companies House Direct) in London to find out if a person is at present a director or has resigned or has been struck off the register. This information is fee-based. Similarly, a report on a company is available online as well.

In Germany databases have also started to supply information on companies online to their members. The same applies to Greece, Belgium, and France.

Databases are not yet available online in Mexico. The investigator in Mexico has to do much legwork by visiting the sources to be consulted, private or governmental.

In Iceland Icecredit provides information from the land register and about the creditworthiness of individuals and companies online for a fee, but full reports are sent by fax or email to the interested clients. They also supply a "national" registry to search for names and addresses of individuals or companies. This search is online and free of charge.

Background checks
In Great Britain a search can be made on an individual quoting an address to see if he is bankrupt. This service is not online. In Germany lower courts publish the names of bad payers and such who have gone bankrupt, but this service is not available online. A justified interest must be proven and a fee is levied.

In Great Britain there is a registry of county court judgments which can be searched by name and address for a small fee. This service is not online. It is similar to the U.S.

In Australia births, deaths, and marriage records are only available to persons concerned who can then apply for the information and collect it, either for themselves or for others who can show a justified interest in the information. Anything collected will be out-of-date three months later. In Australia where no credit checks or police records are available, such searches are mainly done by leg work. The same applies to China and Mexico.

One important part of background checks in the U.S. is the police records on the individuals under investigation. An investigator may register with various police departments. They then do a background check on the investigator before they certify him to do background checks through them. Criminal records are still available from databases such as www.iqdata.com and www.knowx.com.

Locating an applicant's former school or university is of interest to confirm the accuracy of his statement. In the U.S. this search is much easier than in other countries where authorities, schools or universities hide behind the Data Protection Act which is comparable to the Privacy Laws in the U.S.

There are online searches available in the U.S. for national public schools and universities listings: nces.ed.gov/ccdweb/school.asp and www.braintrac.com/.

Any other investigations
Searching newspapers is also a useful tool, e.g. to locate (deceased) people and their relatives when looking at marriage and burial advertisements and stories. The writer of this article once had to confirm the death of a German nobleman in a rushed manner for a life insurance company. The story about the burial event with dignitaries, princes, and high clergymen covered by a local newspaper, together with a photo, provided the proof needed.

There are U.S. newspaper listings by state at the following website: members.tripod.com/~donjohnson/newspapers/states.html as well as international listings. In Great Britain a search can be done through the Press Association to see if there are any press clippings concerning a particular person being investigated. This service has a fee and is not online.

To follow a package sent from A to B may be of importance not only for the sender and recipient, but also in the case of a suspected illicit package, since other parties such as the police may be interested to secure its contents for litigation. Tracing packages sent by DHL, FedEx or UPS, for instance, can be done online www.skipease.com.

In international investigations, to know the time at a certain location or to reconstruct the date of a certain incident, topography of a particular place may be of great importance. Such tools are available online at www.timeanddate.com/worldclock
and www.maps.com/.

Governmental organizations

In Germany there is the project "Online 2005" which aims to have all public entities listed in the Internet by the year 2005, thus augmenting the efficiency of the 650 federal authorities. It is believed that if only half of the procedures required by the citizens are handled online in 2005 (such as change of a marital status), then some 30 billion DM could be saved. The EU Government has similar plans.

In Brombley in Great Britain the "E Government" in miniature already exists to handle all public and other needs of the citizens. In Helsinki, Finland an Internet portal that is a wireless virtual village has been set up, by which any private persons, businessmen, or others can link into the internet of any community.

In Australia police clearance is only available to the person concerned applying for the information and collecting it. The researcher must take into consideration the variations from state to state. Therefore, he needs to have someone on the spot at the time of the inquiry. Furthermore, anything collected now is regarded to be out-of-date three months later.

In Switzerland governmental organizations are available online, and in the U.S. there are abundant links to access governmental organizations or sections of them. Here are some examples:

Florida State:
www2.fdle.state.fl.us.
Florida inmate information:

www.dc.state.fl.us/inmateInfo/InmateInfoMenu.asp
www.dc.state.fl.us/inmate/Release/inmatesearch.asp

California State information:
www.ca.gov/state/portal/myca_homepage.jsp

U.S. state and local government:
www.piperinof.com/state/index.cfm

U.S. federal & county court locator:
www.skipease.com .

New ones are added constantly. Some are free while other are available for a fee. Some may not only restrict themselves to one topic, but supply numerous other searches, e.g. telephone directories, reverse lookups, search for a city by means of an area code or telephone prefix, identification of the owner of a domain name, finding the social security number of a person (the social security number being the key for finding a person in most cases), bankruptcy information. www.skipease.com and www.freeality.com/findet.htm also supply useful links.

The same applies to public records available online: www.investigate-claims.com and a biographical directory of the U.S. Congress: bioguide.congress.gov./biosearch/biosearch.asp to name only a few useful sites.

How fast is online availability in various countries—moving fast, fairly slowly, or not yet started to go online?

In Greece there is a lot of discussion and people look forward to the time when governmental services are available online, probably by late 2002.

As mentioned before, in Germany the process will hopefully be finalized in the year 2005. Some services are already available, such as submitting income tax declaration, and in some states land records may be searched by other authorities online and lawyers can even file suits online.

In Australia things are moving slowly while in Switzerland things are moving quickly. In Israel things are also moving very slowly, and Mexico has a very long way to go before public access is offered to records, and even a longer way to go for computerization.

As you can see, more sources are becoming available online around the world, while others are being shut down due to the Data Protection Act.

I thank colleagues around the world for supplying helpful information to compile this article, and the numerous others who supplied useful websites.

M. Ettisch-Enchelmaier is a Member of the World Association of Detectives/USA (WAD), Lifetime Member of the National Association of Investigative Specialists/USA (NAIS), First honorary member of World Association of Professional Investigators/UK (WAPI), Associated Credit Bureaus/USA, Chamber of Commerce and Industry/Rheinland-Pfalz, Owner of the Investigationsworldwide Association and Ermittlungen-weltweit List, Holder of the Malcolm Thomson Award 1995 and publisher of the monthly "International Business Calendar et Alia." In 1995 M. Ettisch-Enchelmaier was presented with the prestigious Council of International Investigators Malcolm Thomson Award/USA.

Immigration Status Determination
By T. A. Brown

On March 1, 2003 transition of the Immigration and Naturalization Service (INS) was incorporated into the Department of Homeland Security (DHS). Complete Bureau of Citizenship and Immigration Services (BCIS) information for employees can be found online at the official government website at: www.immigration.gov/graphics/index.htm.

One of the easiest ways to determine Naturalization information is via Voters Registration. Most naturalized citizens are proud to vote and therefore register. The registration will show their petition and certificate number as well as the place of issuance.

Check for arrest reports or incident reports. These will also often reflect immigration status.

Does he/she have a good Social Security Number? To verify up to 5 names or Social Security Numbers just call the toll-free number for employers at 800-772-6270 or the general Social Security Administration number at 800-772-1213. Both numbers are open for service weekdays from 7:00 a.m. to 7:00 p.m. in all time zones.

If the alien registration card has a pasted on photo, it's no good. There are several other clues if you handle a card or see a copy of it, but the BCIS only cares if the Permanent Resident Card is properly completed.

Immigrants can now file an online request for Employment Authorization (Form I-765) and a request to renew or replace a Permanent Resident Card (I-90).

If you wish to hire immigrants, you can elect to sign up with the SAVE Program for new hires. It's free to all employers. SAVE stands for "Systematic Alien Verification for Entitlements/Employment." The system is set up through the BCIS. The organization must enter into an access agreement with the BCIS. A prospective employee will provide documentation about their status, and the document number will be entered on the SAVE system which will either confirm the status, or indicate that secondary verification should be initiated. Sometimes status will be confirmed on a secondary verification after the initial request has failed. SAVE online: www.immigration.gov/graphics/services/SAVE.htm

Contact the BCIS Criminal Investigations Division directly to find the date of expiration on a passport or if the immigrant is still in the United States. They search by name, not numbers. The U.S. Census Bureau and Freedom of Information Act requests are also avenues for locating immigration information.

Virtual Immigration Law Library from the Dept. of Justice Executive Office for Immigration Review: www.usdoj.gov/eoir/index.html

BCIS records can be subpoenaed if there is a judgment filed against the immigrant.

U.S. Visa News on Immigration: www.usvisanews.com/

1-800-375-5283 toll-free number for automated case status assistance on Service Center filed cases.

For the Freedom Of Information Act Request Form for BCIS records you need Form G-639 (03-21-94). No prior versions may be used. Other parties requesting nonpublic information about an individual usually must have the consent of that individual on Form G-639 or by an authorizing letter, together with appropriate verification of identity of the record subject. Notarized or sworn declaration is required from a record subject who is a lawful permanent resident or a U.S. Citizen, and for access to certain legalization files.

Freedom of Information Act (FOIA)
Contact: Director
Freedom of Information Act/Privacy Act Program
425 Eye Street, N.W., 2nd Floor
ULLICO Building
Washington, D.C. 20536
(202) 514-1722

Pre-Employee Profiling
By Barbara Klein

I wish I had the ability to profile prospective clients before taking on their cases. I could determine if doing so was going to prove mutually profitable. I wish I could profile a client after the retainer was paid to ascertain if the remainder is going to be paid. I wish I could profile associates so I could tell if we were going to be an asset to each other's working relationship, or a detriment. In my opinion people in investigations and similar professions get "encoding" confused with real life. They reach a point where they cannot tell a falsehood from the truth, kidding themselves and confusing their associates.

In reality, the concept of profiling can be addressed from a variety of angles. My introduction to "profiling" was in college. A branch of the FBI to determine the "profile" of serial killers, mass murderers, etc uses the method. Since the concept was not immediately well received by the FBI, the initial profilers had to prove its validity to gain acceptance. Only after "profiles" proved to be correct over a long span of time did the FBI give it due recognition. The founders of the concept interviewed inmates, looked at the modus operandi of particular criminal activity, and conducted in depth studies to come up with a profile for what type of person would commit a particular crime. Crime scenes were categorized as being "organized" or "disorganized" along with a variety of other factors. The profilers could suggest the suspect would be of a certain race, social status, family background, lifestyle and so, on enabling a more focused search for investigators and law enforcement as a whole.

In the same way a business owner can use profiling to a degree. For example: law firms hire investigators with expertise in this field to "profile" a jury. The investigator is given a list of potential jurors. The investigator then puts together a profile of the jurors enabling attorneys to make more informed selections when choosing and striking people for the jury. Many attorneys do not utilize this very useful service because they think it is too costly. However, attorneys who do hire profilers find doing so has a significant effect on the outcome of a trial. Probably jury profiling will have to earn its way into the industry just as profiling had to earn its acceptance by the FBI.

If one wanted to formulate the overall profile of an individual, one could use a variety of methods to do so. For instance, handwriting analysis can be said to be a form of profiling, as well as body language, kinesic interviewing, and voice inflection. A professional profiler can gain an understanding of the true personality of a person and anticipate how that person may react when confronted with a particular set of circumstances.

If this was so and all of this knowledge was available to us, would it have aided in foretelling or preventing the events of 9/11? Perhaps or perhaps not, since trained, professional "profilers" are not a dime a dozen and are not your run of the mill, John Doe, next door neighbor, common citizen. There are not enough specialists in the field to serve the purposes for which they could be utilized.

Prior to 9/11, while on a cross-country flight, I found the lost ID of someone appearing to look like Osama bin Laden with a script of the native language on the back of the ID. Since I felt the lost ID was of significant importance, I promptly took it to the nearest Security Personnel. From all appearances and judging by the security person's demeanor, I believe the ID was simply trashed. I will never know. When I held the ID in my hand, its importance seemed to be crying out to me, but I naively entrusted it to someone who did not appreciate its significance. Why did I not trust my instincts and pursue the identity of the owner? Was it because my concern was discounted and ignored? Was it because I relied on a person whom I thought would know what to do with this particular ID? I wondered how the person with the lost ID was going to get from Point A to Point B without an ID. It was like an American losing his passport in Europe. Obviously, the security personnel had never been stranded in Europe without a passport.

Had I known, had I foreseen the occurrences of 9/11, how much differently would I have handled the situation?

The point is that although we all seek someone to blame for oversight and negligence regarding the forewarnings of that fateful day, we all individually must take responsibility for that which we can control. What can you and I, as business owners, do to prevent a reoccurrence of disasters of that magnitude or of any magnitude that results in the loss of life? We can utilize background investigations of potential and present employees. If background investigations have not yet been done, start now on prospective and present employees (bearing in mind all the legal requirements as well). Learn more about profiling and how to identify someone who may be a bombshell waiting to explode.

It is not only essential to conduct an initial background investigation, but it is important to conduct subsequent, follow-up checks. Just because an employee is clean today does not mean he or she will be clean tomorrow or the next day or the next month.

There was the case of an actively employed registered nurse who had approximately five convictions for writing false prescriptions. She was serving her time with weekends in jail, and working at the hospital during the week with no one the wiser regarding her activities. In another situation a police officer did not have a driver's license. His license had been suspended and the police department did not even know. When I was doing my internship, a police officer with whom I frequently rode was dismissed shortly afterwards for writing false prescriptions. He was addicted to pain killers prescribed for an on-the-job injury. While I was interning, he would pop pills during the entire shift, and I assumed the fact he was doing so was common knowledge.

Therein lies perhaps one facet of our problem, or at least mine, I confess. I assume too much. I assume someone else will take responsibility for the information I relay. I assume someone is aware of a situation of which I have firsthand knowledge. I assume someone knows, someone else will act, someone else should have, could have, would have...

My personal goal is to take responsibility for that over which I do have control or knowledge. I have learned not to rely on someone else to follow through on what I have initiated or know. It is my responsibility, as a citizen, to be alert to my surroundings, my neighbors, and happenings in my community and business. By doing so, I can help foster a healthier, more secure environment over those areas that I have some modicum of control and perhaps provide a safer environment for those who follow.

As business owners, let us uphold the highest standards possible. Let us do all that we can do to assist, prevent, and deter to the best of our ability events of such tragic proportions. Let us provide a united front with the knowledge we possess. May each one of us assist and provide aid to those who naively walk around with blinders on to the realities of 9/11. Promote security, never stop learning, and fix the things you can that are within your reach.

Barbara Klein is owner and operator of The Profile Agency, a licensed and insured private investigation agency serving northern Alabama. The Profile Agency specializes in background and insurance-related investigations, including profiling.

Active Duty Military Verifications
By Alan Pruitt

A recent change to a Department of Defense database that I have used for two years as a skip trace tool for locating Active Duty military members has made it easier for me to research. Password protection was removed, and this database was moved into a more public domain. Here is the link: bases.defensecity.com. The site is known as SITES (Standard Installation Topic Exchange Service).

SUGGESTIONS

1. Most active duty military member information is either known by the client or revealed in the credit header. Often, an APO or FPO address is associated with this address. Decode the APO or FPO with a resource, such as Debra Knox's "How to Locate Anyone Who Is or Has Been in the Military" (available at amazon.com) to determine the geographical location of the installation.
2. Go to SITES home page and select "View Installation Information."
3. Then select Installation by using one the four search options (Installation Name; Service/Location; Service/Installation Name; or Location/ Installation Name);
4. Select the hyperlink that appears below the installation name that you are researching.
5. Now you have a selection of either "Major Unit Listing" or "Commonly Referenced Numbers."
    a. "Major Unit Listing" will take you straight to a current telephone directory with commercial telephone numbers (and sometimes faxes) for all of the major units at that location. Most of the commercial numbers go directly to the Commanding Officer or the personnel section that knows how to contact your military member target. NO PRETEXT IS REQUIRED! Most information about active duty military members is public record and can be easily verified with whoever answers the telephone. Most clients want to know the "duty section" telephone of the member, as well as the rank and name of the CO (Commanding Officer). It's free information so just ask!
    b. "Commonly Referenced Numbers" will sometimes list a "Base Locator" commercial telephone number. Sometimes yes; sometimes no. It depends on the

installation. Again, NO PRETEXT IS REQUIRED! Just ask if the member is listed and what is the unit assignment and direct commercial telephone number.

c. Sometimes you can call after regular duty hours and have a low-ranking clerk assist you by actually pulling up records on his or her computer or pulling a file from a filing cabinet and verifying information for you over the telephone. Then, when you call during regular duty hours, you are better prepared to navigate the military machine. Get names of commanders or officers at every turn. It helps to melt through any stonewalling.

d. Click on "Must Know Items" and you will find the actual mailing address and physical delivery address for the installation. This is vital to repossession companies.

VERY few active duty military members have classified or "secret" units or locations, so don't buy that load from a well-intentioned military clerk at the other end. Just jump the chain until you get results or call other adjacent units and steer in from a different direction. Again, I have never had to use a pretext in any of my military locates. Just ask for the information!

International Driver's Licenses
By T. A. Brown

If you are traveling out of the country for business and find yourself needing to drive to a location, this information will make it a quick and simple process to obtain an international permit to do so.

The World Convention established International Driver's Licenses in 1949. You can get an International Driving Permit at any AAA office around the country. It is a multiple page booklet that basically just translates your driver's license into a half dozen of the most popularly spoken languages. Just take in your current Driver's License and a photo. It costs is about $7. Although these International Licenses are good everywhere, except in the country of issuance, they are not required for you to drive a vehicle in another country. All you need is a current and valid driver's license and your passport. The International Driving License is simply intended to let Police Officers in other countries know that you have a regular license because of the many languages printed on it. Not all countries recognize them, so you still must have a valid driver's license from your place of origin. The unofficial exception is that intelligence personnel stationed overseas for civil purposes will always drive with an international permit in case they get involved in an accident.

The holder of an international driver's license may drive in a foreign country for one year, or until the expiration of the permit. If a foreign tourist comes to the United States with an international driver's license and becomes employed or enrolls in school as a student, they must apply for a local license in the state where they take up residence. There are no exceptions. If a potential employee produces an international driver's license for identification and is looking for work in the United States, they cannot be legally hired unless they apply for and obtain a state driver's license.

The international driver's licenses that are being sold over the Internet are questionable and are being investigated by the federal government. To be safe, get yours through the AAA or your local auto club.

Background and Credit Investigations
By Charles T. Rahn

This is the form needed for employment that meets the FCRA rules. If you are not aware of the FCRA rules, then you need to make sure that you are brought up to date. There have been many changes made and you need to become familar with them. You can use this to your advantage. Copy the forms out and send them to the directors of human resources in your area. Many H.R. directors will not be aware of these changes.

I state that the information I have provided to

_____

with regards to my seeking employment is true and complete. I understand that any false statement(s) made in this regard will result in my not being offered employment or in termination of my employment. I further understand that this authorization is not and is not intended to be a contract of employment, nor does this obligate _____
in any way if it determines not to employ/train me.

I hereby authorize _____ or any of its agents to make an inquiry into my personal history, education, employment, credit records, driving records, and criminal history through any investigative or credit agencies or bureaus.

_____ may request such reports for any purpose it deems appropriate, including, but not limited to, inquiries permitted by law.

Dated: _____

_____
Name of applicant (last, first, mi)

_____
Signature of applicant

_____
Social security number

_____
Drivers license number/state

Date of birth

The full text of the FCRA as amended may be found on the Internet at: www.ftc.gov/os/statutes/fcrajump.htm.

I strongly suggest you read the amended rules. They now affect surveillance on employees. If you as an employer have one of the new forms signed and on file, then and only then can you cause a surveillance to be done on that employee. If you do a surveillance and this form is not on file, then both the surveillance person and the employer can be sued.

Chapter Three

Workplace

Preventing Workplace Violence: Management Considerations

By Robert A. Gardner, CPP

It's ten in the morning and the sounds of a disturbance in your outer office distract you from the file you've been studying. Suddenly your office door bursts open. Standing in the doorway is Charlie Smith.

Until last week Charlie worked in accounting. He'd been a model employee for more than six years, but recently he began to change. He became withdrawn and moody. His work deteriorated, and he was prone to violent verbal assaults against co-workers. After he tried to attack his supervisor during a counseling session, you were forced to terminate him.

Annoyed by the disruption, you rise from your desk. You start to protest the intrusion, then suddenly stop. A wave of terror sweeps through you. Charlie has a gun!

For most of us, workplace violence is something that only happens on the six o'clock news. We know that it exists, but there is an air of unreality about it. We can't relate to mass murder committed by a deranged postal worker or to the scattered wreckage of a commuter airplane by a disgruntled former airline employee. Incidents like these are so far from the realm of our everyday experience that we just can't comprehend them. As a result, we tend to view the idea of workplace violence as a sensational curiosity. It happens, but not in a business like ours and certainly not to people like us.

Unfortunately, workplace violence isn't limited to the occasional murder rampage in a government building. It can and does happen anywhere. Every business, regardless of its size and type, should have a workplace violence program in place. This program should:

- Understand the consequences of workplace violence incidents
- Organize a crisis management team
- Develop and disseminate a workplace violence policy
- Identify and evaluate potential threats of violence
- Provide mechanisms for reducing or eliminating threats

The Consequences

Violence impacts organizations in a variety of ways. The tragedy of dead or seriously injured employees is obvious. Less obvious is the damage an organization can sustain from the consequences of actual or threatened violence. Morale and productivity suffer when employees are frightened or disturbed by violent incidents or threats of violence. Should the victim be a key employee or important executive, day-to-day operations can be seriously disrupted. Employees who lose confidence in the organization to provide them with a safe working environment may be inclined to look elsewhere for work. Fear and anxiety can induce a variety of personnel problems that can sap the strength of the organization.

News of violence within an organization can also become a public relations issue. The negative attention brought on by a violent incident can have far-reaching ramifications. Customers may be frightened away. Associates may sever ties to avoid the risk of their names being linked to the violence. Recruiting efforts can suffer. Prospective employees are likely to avoid an organization with a violent reputation.

Liability is also an issue. It is virtually certain that any violent incident will result in some sort of litigation. It is also certain that the litigation will attempt to show that the organization was somehow negligent in its approach to workplace violence.

In cases where an employee is the cause of injury to others, there will inevitably be claims that the organization knew or should have known of that individual's propensity for violence and failed to act. Issues of negligent hiring, negligent retention, and inadequate supervision will be raised. Individual managers and supervisors will be singled out for their alleged negligence in failing to predict and prevent the violence. In situations where outsiders injure employees or customers, the organization will be accused of negligence for failing to provide adequate security. Claims will be made that the violence, whether caused by employee or outsider, was foreseeable and but for negligent and inept management practices could have been prevented. As with other security and safety considerations, having a formal workplace violence program in place can go a long way toward mounting a credible defense should claims of negligence be made.

Crisis Management Team

Formation of a Crisis Management Team is the first essential step in the development of a successful workplace violence program. Effective management of a workplace violence program requires input from a variety of disciplines. No single individual has the training, experience, and authority needed to develop, implement, and administer the necessary policies and procedures.

The ideal Crisis Management Team should consist of a member of executive management along with a management member from human resources, security, and risk management. There should also be a legal advisor and a psychologist. Some organizations may also choose to include an employee representative. This group should be charged with developing workplace violence policies and should be the decision making body when workplace violence issues arise or incidents occur. The Crisis Management Team should report directly to the organization's highest executive officer.

Frequently organizations do not have the personnel needed to adequately staff a Crisis Management Team. In those cases it may be necessary to retain outside experts to fill some functions. It is common for organizations to operate without a security department, an in-house attorney, or a staff psychologist. Private practice psychologists, private legal counsel, and independent security consultants can be called in to fill these rolls.

Regardless of the source of the team membership, it is critical that they have the full and active support of executive management. The executive management member should be the team leader and must have sufficient influence and authority to overcome any organizational obstacles to the successful operation of the team.

Individual team members should have a working knowledge of the organization as a whole and should be experienced experts in their particular discipline. They must be able to fully comprehend the complex issues surrounding workplace violence and be prepared to offer viable solutions to problems affecting their particular area of responsibility.

It is important that the team meet regularly. Once the workplace violence program is developed and in place, it must undergo continuous review. Regular meetings permit the team to keep current on the daily operations of the organization and assist them in identifying and evaluating potential problems. Activation of the team only after problems occur puts them at a serious disadvantage in maintaining an effective program. A proactive approach not only aids the team in monitoring and managing the program, but it also puts the organization in a much better position to defend itself should an incident occur and litigation result.

The team should also be tasked with contingency planning. Should an incident occur, there must be a response plan in place to deal with it. Trying to develop a plan under fire virtually guarantees failure. Every possible scenario should be considered and a response for each developed and tested.

Consideration should be given to making the Crisis Management Team the nucleus of the organization's overall emergency preparedness effort. Members of the team should occupy key positions in the organization's critical incident command system. Workplace violence incidents should be fully incorporated into the organization's disaster preparedness program with periodic drills conducted to ensure that the system works. In reality, workplace violence is simply a different form of disaster. Many of the response procedures are similar to procedures for dealing with other workplace emergencies.

In the formative stages of a Crisis Management program, it may be helpful to retain the services of an independent Crisis Management expert to act as an advisor and facilitator. This individual can provide insight and guidance to the team and act as a technical resource for the development of policies and procedures.

Policies and Procedures

The foundation of a Crisis Management program is its policies and procedures. These should be carefully researched, easily understood, and committed to writing. In practice there should be a number of separate but related policies in place. Each policy is geared toward its particular target audience and contains the information that audience needs to function within the program.

An example would be a Workplace Violence section in the organization's personnel manual. Although it would seem to be common sense that violent acts are prohibited in the workplace, a specific rule to that effect would remove any doubts and make imposing discipline easier when the need arises.

The manual should describe procedures to be followed by employees who are victims of actual or threatened violence. Reporting procedures should be in place for employees who witness violent acts or who are concerned that a violent situation may develop. The manual should identify any employee Assistance Programs that may be available and provide instructions for accessing these programs.

Contingency plans describing procedures to follow should be prepared and distributed to team members and other personnel who may be called on to make decisions during an actual incident.

These plans should include operational checklists, emergency contact lists, notification schedules and resource lists backed up by detailed operational plans. Unlike personnel manuals that are given to everyone, contingency plans should be treated as confidential documents. They contain information and action plans that would be useful to anyone contemplating violence against specific individuals or the organization in general. Limit their dissemination only to those with a need to know.

Identifying Threats

Shock, surprise, and disbelief are common reactions to workplace violence incidents. However, analysis of individual cases often shows that there were warning signs that went unrecognized or were ignored.

As part of the Workplace Violence program, organizations should train employees at all levels to recognize and report the danger signals that often precede incidents of violence. Each of these reports should be evaluated by the Crisis Management Team to determine if further action is warranted.

Because of the sensitive issues involved, great care must be taken to ensure that the rights and the privacy of everyone involved are closely guarded. An employee who exhibits "at risk" behaviors must be evaluated, but they must not be made to feel they are being singled out for ridicule or persecution.

Predicting human behavior is an inexact science at best, but experience has shown that individuals who commit violent acts at work often fit one or more of several "at risk" profiles. There are no absolutes when classifying "at risk" employees, but a red flag should be raised when an employee:

- Exhibits emotional instability or violent behavior
- Exhibits signs of extreme stress
- Undergoes profound personality changes
- Feels victimized by supervisors or the entire organization
- Makes threats or alludes to acts of workplace violence
- Exhibits signs of extreme paranoia or depression
- Displays behavior inappropriate to the situation at hand
- Exhibits signs of drug or alcohol abuse
- Is involved in a troubled, work-related romantic situation

When an employee is thought to be "at risk," the organization must take immediate action. Responsibility for evaluating the risk and developing a response plan should be coordinated through the Crisis Management Team.

Along with "at risk" profiles, the organization should be sensitive to "trigger situations." These are events that can serve as a catalyst to push a violence-prone employee over the edge. Normal, emotionally stable employees may show little or no reaction to "trigger situations." If they do react, it is usually in a controlled and reasonable manner. The "at risk" employee, on the other hand, may view "trigger situations" as events that justify a violent response. It would be impossible to list every conceivable "trigger situation," but there are some events that are common to the workplace and should always be viewed as potentially dangerous. These include:

- Performance counseling sessions
- Disciplinary actions
- Termination, including non-disciplinary lay-off
- Non-selection for promotion or a desired position
- Criticism or harassment from coworkers
- Failed or spurned work-related romance
- Significant non-work-related personal crisis

Careful management of these and other "trigger situations" is an essential element of a Workplace Violence program. In many of the above situations, the organization has considerable control over the conditions under which interaction with the employee occurs. Through preplanning it is possible to exercise this control in a manner that ensures that the possibility of a violent reaction will be markedly reduced.

Threat Reduction
The ultimate goal of any Workplace Violence Program should be the total elimination of all work-related incidents of violence. This goal may not be entirely achievable, but an effective program can significantly reduce the risks. The first step in preventing violence is awareness by management that a potential threat exists. In order to achieve that awareness there must be a mechanism for all employees at all levels to report potentially violent people or situations.

The ideal system should bypass the normal chain of command and report directly to a designated member of the Crisis Management Team. This centralized reporting ensures that the information will reach the decision makers quickly and serves to facilitate the identification of "at risk" behavior patterns. A troubled employee may make threats or display other symptoms in several locations throughout the organization. Centralized reporting enables the Crisis Management Team to assemble data from separate sources and recognize problems far sooner than possible with conventional upward reporting. Centralized reporting also depersonalizes the reporting process somewhat. It can encourage reporting in circumstances in which the employee may be reluctant to confide in a supervisor whom they deal closely with on a daily basis. This is particularly true when all involved parties work in the same department or area.

Once identified, "at risk" employees must be evaluated and, where appropriate, provided with counseling or other help. In many cases this can be accomplished by referring the employee to the organization's Employee Assistance Program. If no such program exists, arrangements should be made to have the employee see a company retained psychologist. Even in situations where the employee's conduct clearly warrants termination, an evaluation should be conducted. This could preclude—or at least provide warning of—possible violence during or after the termination.

The availability of counseling through an Employee Assistance Program may also encourage an employee who recognizes that they are "at risk" to seek help before a crisis occurs. This opportunity to get help without fear of discipline or condemnation saves both the employee and the organization from the consequences of a tragic confrontation.

Screening Out Problems

During post-incident investigations, it is common to find the offending employee had a long history of violent or disruptive behavior. Had that fact been known prior to employment, they would almost certainly not have been hired.

Pre-employment screening, sometimes referred to as a background investigation, should be an integral part of every organization's personnel policy. Unfortunately, few organizations conduct thorough investigations. This failure not only subjects the organization and its employees to unnecessary danger, but it also exposes the organization to serious liability. Lawsuits involving claims of negligent hiring are among the fastest growing areas of civil litigation. Properly designed and implemented pre-employment screening programs can quickly and legally identify unsuitable applicants, thereby shielding the organization from a multitude of dangers.

Because we live in an era that emphasizes the rights of the employee over the rights of the employer, many organizations feel that conducting anything more than a cursory reference check—or conversely, providing anything more than dates of employment in response to a reference check—is somehow illegal. Providing that applicable civil rights laws are observed and the applicant has signed a waiver permitting it, there is nothing improper or illegal about making an extensive background inquiry.

A background investigation based solely on the information entered on an application form by the prospective employee is a waste of time. The information on the form is important to an investigator, but applications are routinely falsified. Non-existent job histories are created; breaks in employment glossed over, and education exaggerated. Because applicants rely on the fact that employers tend to accept application data at face value, far too many applications are works of fiction rather than a true picture of the applicant's history.

In addition to background investigations, employers should consider the use of pre-employment testing to screen job candidates. There are a number of approved tests on the market which can assess a wide variety of character traits including honesty, substance abuse and propensity to violence. The implementation of an effective pre-employment screening program is essential not only to workplace violence prevention, but also to the overall well being of the organization.

Violence From Outside

There are two primary outside sources for violence against the organization and its employees. The first is criminal in nature. This type of violence has no close personal connection to the organization. Criminal attacks are usually carried out by strangers and are motivated by either economic factors or ideology. Typical examples would be robbery or terrorism. Prevention of these and similar crimes requires an effective physical security program. Such a program integrates facility design, security hardware and electronics, security personnel, organizational policy, and local law enforcement.

The second source of outside violence has a more personal connection to the organization. These attacks typically involve former employees or non-employees with an emotional connection to the organization. An example would be a terminated employee with a grudge against the organization or against an employee of the organization. Another example would be a jealous or rejected lover who blames the organization or someone in it for their problems. An unstable husband who knows or at least suspects his wife is having an affair with a co-worker could pose a serious threat.

In some ways the criminal threat and the personal threat are similar. Both can require that the attacker circumvent security precautions and both can result in serious harm if not prevented.

They are different, however, by virtue of the fact that the personal attacker may have access to the organization that a stranger is denied.

There is also the added dimension that the organization may have a responsibility to provide protection to a threatened employee away from the workplace. An organization usually has no responsibility to protect an off-duty employee from criminal attack. However, a company executive, or any other employee for that matter, who is the subject of work-related death threats could be entitled to receive around the clock protection. Failures to provide adequate security for such an individual could result in liability for the organization should an incident occur.

When faced with providing protection against disgruntled former employees or other dangerous outsiders, the organization should look to the law for whatever help may be available. Under some circumstances police may be able to intervene immediately. In most cases, however, the mere threat of violence does not give police the authority to act. In California an anti-stalking law can be applied to some workplace violence situations.

Also available are Temporary Restraining Orders. These orders are issued by a judge to prevent individuals from committing certain specified acts. In workplace violence situations these orders can sometimes be useful to prevent harassment of employees. Violation of the provisions of a Temporary Restraining Order is a crime. This can give police the authority to act even when no other crime has been committed.

The Choice

There is no guaranteed prevention program for workplace violence. Human nature is too unpredictable for that. There are however proven techniques for minimizing the risk of workplace violence by recognizing its danger signals and acting upon them.

Organizations that accept the possibility of workplace violence and actively plan to prevent it stand an excellent chance of avoiding a violent incident. Organizations that choose to ignore the danger run the very real risk of becoming the lead story on the six o'clock news.

Robert A. Gardner, CPP, is a former Corporate Security Manager and Police Crime Prevention Specialist with over 32 years experience in security and crime prevention management. His expertise includes business, personal, housing security, and architectural/environmental design security. He has provided technical guidance and training to property managers, developers, homeowners, business managers, employees, and others in the recognition and reduction of security risks and the implementation of security and crime prevention measures.

He has also served as an advisor on architectural/environmental security to a City Planning Department, developed that city's business crime prevention program, wrote its Security Alarm Ordinance, managed its Building Security Ordinance program, and served as Assistant City Emergency Services Coordinator and Media Relations Officer.

In the private sector he worked as Security Supervisor for the West Coast headquarters of a national retailer and as a Regional Security Manager, Store Loss Prevention Manager, and Electronic Security Specialist for a major hardware / department store chain.

He has been designated a Certified Protection Professional (CPP) by the American Society for Industrial Security and a Certified Security Professional (CSP) by the California Association of Licensed Investigators. He is a graduate of the California Crime Prevention Institute and its courses in Advanced Crime Prevention Through Environmental Design, Earthquake and Terrorism, and California State Police Executive Protection.

Note: Additional Information on Workplace Violence Prevention can be found on OSHA's Home Page: www.osha.gov/

K-9 Use in the Workplace
By Justin Spence

Many people, when thinking of a police service dog (otherwise known as a "K-9"), might think of a large, snarling German Shepherd with a bad attitude that only wants to do one thing: attack...attack...attack.... This is the farthest thing from the truth, and due to these misconceptions many agencies and businesses alike are missing out on the many uses of a K-9.

In fact, police service dogs are even-tempered, sociable animals that do far, far more than suspect apprehension and handler protection. These specifically trained animals must be able to play with and love a child one moment and be able to apprehend a fleeing felon the next—and go from one extreme to the other with ease. They only make the cut and are able to become certified if they have those qualities that keeps an ill-tempered dog from ever obtaining a position as a police service dog.

Among other specialties, detection K-9s are specifically trained to detect different substances, such as narcotics, alcoholic beverages, weapons, accelerants, and explosives.

K-9 Uses in Business Environments

Unexplained inventory shortages, decreased production, rising injury rates, damaged products, and a host of other issues can signal hidden problems in the organization. Problems such as employee dishonesty, drug and alcohol abuse, and fraudulent workers compensation claims can add to financial ruin for any business.

Likewise, if your company has industrial accidents or fatalities caused by on the job drug use, you will inevitably experience the legal and moral ramifications that follow.

Studies by the Labor and Resources Subcommittee on Employment and Productivity found that, compared to the average worker, drug users are absent from work 16 times more often, have an accident rate four times greater, use over 33% more sickness benefits, and file five times more compensation claims. Another study found that 47% of on the job injuries were directly related to drug/alcohol abuse. Furthermore, it is estimated that over $60 to $100 billion is lost annually in productivity by U.S. corporations and small businesses.

The Federal Drug Free Workplace Act of 1988 requires that contractors and grantees provide a safe, drug-free environment for their employees.

Specifically trained detection canines can quietly and quickly check for just about any type of illegal contraband in virtually all environments. This can aid in fire investigations to check for accelerants, narcotics investigations (in schools, businesses, or even for concerned parents) to check for many types of illegal narcotics, or for workplace investigations to check for weapons or even alcoholic beverage consumption on the job site. They are also very useful in detecting explosives for schools, businesses, or executive protection contracts.

The detection K-9 can pick up on odors 200+ times better than a human can, thus making them an invaluable resource. A detection K-9 can sniff a set of time cards and will be able to detect the user from that sniff. A detection K-9 can sniff out a single bullet in a briefcase or locker.

A K-9 trained on the "passive" alert method (in which they simply sit when they detect the odor they are trained on) can even perform sweeps on an employee or student without them even knowing, by just walking the dog past them. Although this type of search does fall under the Fourth Amendment and a person's right to privacy, it is not advised to do this type of procedure unless you are trying to secretly weed out who may be the user in the group. This passive method is the preferred method of response for use in school or business environments, since it does not raise anxiety levels and is a quiet alert that does not cause destruction to property.

A detector dog sniff, although sometimes called a "search" or "sweep," is not a search as defined in the Fourth Amendment. The Fourth Amendment covers people, not places, and therefore anything in public and semi-public places is subject to a detector dog sniff. Such things are lockers, hallways, vehicles, trash, etc.

Some useful Supreme Court rulings that back up detector dog sniffs are:

Doe v. Renfrow, The case allowed a subject's detainment in order to obtain a detector dog to sniff for narcotics in the subject's luggage. It was not classified as a search and did not infringe upon the subject's Fourth Amendment rights.

Horton v. Goose Creek Ind. School District, The case showed that schools are public places and that when the lockers, hallways, student's backpacks, desks, and vehicles were subjected to a detector dog sniff, it was not a search.

Places covered by the Fourth Amendment:

1. Private property
2. Vehicles
3. Homes
4. The curtilage of homes
5. A person's body

Places not covered by the Fourth Amendment:

1. Abandoned property
2. Public places
3. Trash once it has been abandoned
4. Open fields
5. Things exposed to public view

Myth: A drug detection K-9 cannot detect the narcotic odor if covered up by coffee, gas, grease, or other harsh masking odors, etc. "I have had dogs alert on a truck with dope hidden in a false gas tank, covered with gas and dipped in engine grease. That is why they are trained with proofing odors, which include anything from dog food to perfume to tennis balls, etc." JS

Justin Spence is an expert in Explosives Investigations and Domestic Terrorism Preparedness. He has trained thousands of federal employees in basic explosives identification and detection and in the proper use of Explosives Detection Machines and Trace Detectors in an airport environment. He heads up the K-9 Support Unit of Source Investigations Group, Inc., based in Orlando, FL, that provides support for special investigations that involve narcotics, explosives, or may need a higher level of protection.

Avoiding Computer Lockups
By Kevin J. Ripa

Note: The same caveat applies to this technical article as the Computer Security and Recovery article. Please read disclaimer on Copyright page before proceeding.

Computer lockups can typically be classified into two categories: Hardware Lock Ups & Software Lock Ups.

Hardware is defined as any physical computer gunk that you can actually hold in your hand. A monitor, mouse, keyboard, mother board, RAM, hard drive, sound card, modem, etc. are all examples of hardware. Software is defined as an intangible because there is no physical evidence of it other than the disk it came on. Software is the program itself. This can include the program to make your modem run. Windows 98, Outlook Express, Microsoft Word, Eudora Pro, and computer games are all examples of software.

Because there are many different types of hardware, it is very difficult to get into a short discussion about it. Besides the physical hardware, you must consider things like IRQ assignment, DMA Controllers, PCI Controllers, USB, blah, blah, blah. Fortunately, hardware lockups usually account for less than 1% of computer lockups.

Software Lockups
These are by far, the most common lockup you will see. It works on the idea that you are asking the computer to do two things at once and each of these things needs the same resource to do it. They fight over the one resource available, and while they are fighting, nothing will happen. This will cause a lockup at the very least and the BLUE SCREEN OF DEATH at the worst. Fortunately, these are easy to fix since you caused the lockup in the first place. Let me explain. Have you ever been working away and all of a sudden your computer locks up? Usually just as you have tried to open a program? Ever tried to open a program and you get an error message saying something like "Could not open such-and-such. .DLL whassisname could not be found."

These .DLL thingies are terribly important. .DLL is a file extension that stands for Dynamic Link Library. These are little files common to many different programs that they need to open and run. (One of many). Anyway, back in the days of 386/486, etc, when hard drive space was at a premium, programmers decided that if a bunch of different programs used the exact same file (i.e. 123.DLL) to work, it would be a waste of precious space to duplicate this file for every program that needed it. They decided that if they could just have one copy of 123.DLL, all the programs that needed it could share, and it would save space. This was a really good idea, because now there is one copy of 123.DLL and whenever a program needs this file to operate, it will create a duplicate of the file for the time it needs it and then throw it away when it is done.

When you try to delete a program without the uninstall program usually provided with the program, in other words, going into your explorer and just dragging that file to the Trash bin because you don't need it anymore, it takes whatever .DLL files are associated with it and deletes them too. And I am talking about that main .DLL file. So now it is history. When you try to open another program that needed that .DLL, it won't work because you threw it in the trash. Unfortunately, even though the .DLL is a shared file, each program thinks the .DLL is part of itself and throws it away. It isn't smart enough to know that it should leave that .DLL file behind. That is why uninstall programs are so important. They know how to separate the .DLL files and leave them behind while throwing the rest out.

So where are these uninstall programs? If you click on Start/Programs and then move your cursor over the specific program you want to uninstall, a menu pops up showing all of the options to that program (either opening it, help file, etc). Listed there should be an uninstall icon. Click on it and follow the instructions. Make sure before you uninstall it that all other programs are closed! Let me jump off track for a moment and explain this.

The first and **MOST IMPORTANT** thing to do is to close **ALL OTHER** programs that are running in the background. This applies to all installs and uninstalls of any programs **EXCEPT** for users of Windows NT, 2000, and XP. Now you may think that they are all closed because since you turned your computer on, you haven't opened anything. **WRONG** (usually). Press Ctrl+Alt+Delete keys all at the same time. You will see a window pop up called the Close Program window. Let me reiterate! Windows NT, 2000, and XP users can ignore these instructions. They do not apply to your system.

Back to the show. Everything listed inside there is a program that is running in the background for one reason or another and was activated when you started your computer. No matter what you see in the Close Program window, highlight it and click on End Task. You can only do them one at a time and then you have to press Ctrl+Alt+Delete to make the window come up again. Keep doing this until all that remains is EXPLORER and whatever the install program listing is. You should be able to recognize it. **DO NOT** close Explorer because if you do, your computer will stop working! Explorer is the program that gets the necessary files in the right order to make the computer do what you want it to do! Some of the programs take a bit of time to close so don't rush them. End Task one and then wait a couple of seconds. The one labeled Systray (your system tray on the bottom right hand corner of the screen) is like this. After a few seconds you may get a box that pops up and says something like "Program not responding." Then it will ask you if you want to wait or shut the program down. Click to close program. Now, when you are left with only Explorer and the install program, proceed with the install. If you are not too proficient on a computer, once **ANY** install is done print off the README file. When you are done installing, you should automatically be asked if you want to do this. I would suggest that 60% of the questions related to programs could be answered by reading this. This file can also include late breaking bug fixes with the program that were found after the program was released.

Now, restart your computer. **VERY IMPORTANT!!!!** A system restart should be done anytime you install or uninstall any program, even if the computer does not prompt you to do this! The reason is that when you install or uninstall, funky things happen in your computer. I'm not even going to get into that here. Suffice to say that a restart lets Windows "sort" itself out and puts everything in the right place. Doing this will also save you from having to Defrag your computer as often. Everyone knows what a defrag is right??

Back to DLL thingies. If you remember, I said earlier that I would explain why you have to uninstall the proper way. Well, here it is. If you uninstall a program while another one is running that shares the same files (or even when it is not), it can either steal the file from another and throw it away or NOT delete files in the uninstalling program that it should be uninstalling and just generally gets confused and causes problems. This is one of the causes of your computer running slowly and generally frustrating you.

Having said all of this, people will jump up and down saying, "Well I never shut my programs down when I install/uninstall and I have never had a problem."

True. Could happen. Usually happens. Works okay nine times out of ten times. But it is that one time that you are going to be really choked when you lose data because of a crash or lockup. Your potential for problems increases exponentially when installing or uninstalling something as intrusive as AntiVirus software and Firewalls.

Let us look at it at the opposite end. Consider when you install programs while other ones are open. If a .DLL needed by the program being installed is already being used by another program, it does not create the appropriate file path to the .DLL. What happens is that if you have the one program open and you go to open the other program, the first program will be using the .DLL and the second program won't be able to duplicate it for use. It will want the .DLL all to itself. Imagine not being able to slice a hot dog in half so the kids can share. And you don't have any more hot dogs. The two will fight over the hot dog and no one gets to eat while the fighting is going on. Same as with the .DLL. The two programs will fight over the .DLL in a kind of "Boolean" (that's computer language) duel and neither will work. Hence, computer lockup. When this happens and you have hit all the keys and nothing is happening, (you know what I am talking about), hit Ctrl+Alt+Delete (this command SHOULD work). The program causing the problem will be listed, and in brackets beside it it will say Not Responding. Highlight this and click on End Task. The locked program should stop running. Now restart your computer to clean things up.

What if not all the programs in your program menu have an uninstall icon? Well, in that case, try this. Click on Start/Settings/Control Panel/AddRemove Programs. Do you see the program listed there? If yes, then select it and click remove. All the same rules apply as far as closing all running programs.

Not listed there? Consult the Help file of the program or check the manual or email the program's tech support and they will tell you how.

Now you know why your computer locks up and how to avoid these lockups in the future.

Here is a brief summary. **ALWAYS** use the program's uninstall program to remove it. **ALWAYS** make sure that all programs **EXCEPT** Explorer are closed via the Close Program Box **BEFORE** installing or uninstalling **ANYTHING**.

Now I am going to explain how to clean up your computer without doing a reformat. **GO OUT AND BUY NORTON SYSTEMWORKS**. When you install it, you will be prompted about the programs within the SystemWorks that you want to install. Just install the utilities and the anti virus. Utilities will now, upon your instructions at any time, scan your entire computer (hardware and software) and find any and all problems in your computer. This includes missing .DLLs, bad file paths, bad application data, and all the things that cause your computer to run slowly or lock up. The best part is that it also FIXES all of these problems with the click of a button. VERY easy to use.

I am not a big fan of freeware in any capacity. The more little programs that you have on your computer, the more problems you are asking for. Unless you know exactly how to configure the program and exactly what it does, it could cause more problems than not, because in most cases these little free programs are not integrated into a Microsoft environment. For this reason they can be very unpredictable. Microsoft had a lot of problems in the early stages of IE5 because of all of the peripheral programs that it included with its browser. So only download what you NEED, and understand it before you download it and know how to configure it BEFORE you install it.

One last tip. If you hear the hard drive grinding away, wait until it stops to do anything else. This will allow the computer to close or open or save programs one at a time. This goes for closing open windows as well. Slow down!!! This little box is no different than us. If we do one thing at a time, we can do it well, give us five things to do at a time and they may all get done, but not as well as one at a time.

It is my sincere hope that these explanations were presented in a common sense, easy to understand manner. I hope that you can see that computer lockups and instability is not the magical, mystical beast that many would have you believe. Clean, smooth computing can be exercised by anyone just following a few simple steps. Should you ever have any questions, you can contact me by email at info@computerevidencerecovery.com or visit my website at www.ComputerEvidenceRecovery.com.

Domestic Violence: A Concern for Employers?

By Coleen Widell & Rick Naylor

Kenneth McMurray, Chad Anderson, Richard Farley, James Simpson, Charles Lee White, Dean Carter, and James Calvert. These men are computer analysts, factory workers, technicians, grocery clerks, laborers, and police officers. What do they have in common? They entered the worksite of their wives and girlfriends ...and murdered them (and/or others) at work.

Employers Are Concerned

In 1994, Liz Clairborne, Inc. commissioned a survey of corporate security leaders entitled, "Addressing Domestic Violence: A Corporate Response." This survey involved interviews with senior executives in Fortune 1,000 companies in the United States. There were several key findings of the survey:

- 94% of corporate security and safety directors ranked domestic violence as a high security problem.
- 33% said domestic violence affects their balance sheets.
- 50% of the executives say domestic violence has had a harmful effect on their company's productivity, attendance, and health care costs.
- 66% agree that a company's financial performance would benefit from addressing the issue of domestic violence among its employees.
- 40% are personally aware of employees in their company who have been affected by domestic violence.
- 90% were aware of *more than three* incidents in which men stalked women employees.

Domestic violence in the workplace is the newest security issue for this millennium. Today's most effective HR and Security directors are addressing the impact of battered workers to reduce their legal exposure.

Legal Obligations

Federal legislation is very clear regarding employer obligations for creating and maintaining a safe workplace. Simply put, employers have the duty to provide a relatively safe workplace for their employees. This duty has been created in several federal or state statutes, regulations, or case law.

Businesses must become familiar with these legal requirements to reduce the impacts of liability for workplace violence. Seven-figure lawsuits are not uncommon. Enormous legal judgments have been entered against employers who have failed to protect their employees when domestic violence comes to work.

The Extent Of Domestic Violence
This epidemic of domestic violence is far reaching indeed. Conservative estimates reveal between 3 and 4 million women are physically beaten every year in America, averaging an assault every *9 seconds*. [By the time you finish reading this article, another fourteen women will have been beaten.] Domestic violence is one of the most pervasive, yet under-reported crimes in America. Look at the women you know in your life: your wife or girlfriend, your mother, sister or niece, your co-worker, gym partner, neighbor, yourself (?). It is likely at least one or more of these women either are or have been victims of intimate violence.

Since almost half of the workforce in this country is made up of women, a battered woman is most often a working woman. In fact, 70% of domestic violence victims are employed.

The Impact On The Workplace
What is the likelihood of this happening in YOUR workplace? National statistics tell us the effects of domestic violence in the American workplace are far reaching, indeed. Of battered workers, 74% were harassed at work by the batterer, and 96% of battered workers experience problems at work due to abuse by a partner.

Domestic violence becomes a workplace safety issue when a batterer makes threats or comes to the workplace seeking to harm and intimidate his victim. In the United States, current or former husbands or boyfriends commit 13,000 acts of physical violence against women in the workplace every year, and these men kill 17% of the women killed at work.

In addition to the likelihood of violence spilling over into the workplace, employers often endure a milieu of other negative effects generated by the battered worker's domestic situation. For example, over 50% of battered workers suffer from tardiness and absenteeism problems. Lost productivity—from both the battered worker and her co-workers who indirectly experience the violence—is not uncommon. So too are:

- Poor job performance
- Increased use of sick days

- Reduced quality
- Decreased customer service
- Employee turnover (requiring costly rehiring or retraining)
- Increased health insurance premiums

The Corporate Cost

You are being financially assaulted by a milieu of "hidden" costs that you probably have not yet detected. As an employer, you are incurring the expense of "domestic violence in the workplace." You pay the bill, but you have no idea why the cost of doing business continues to skyrocket. Simply stated, the effects of domestic violence are eating away at your bottom line. Your profits are being diverted into "business expenses" directly related to the corporate cost of domestic violence.

The Bureau of National Affairs estimates the corporate cost of domestic violence to be between 3 to 5 billion dollars annually—for medical expenses and increased health care costs directly related to domestic violence. In addition to that, businesses forfeit another $100 million in lost wages.

Businesses who have been victimized by domestic violence in the workplace also face increased workers compensation claims to the tune of approximately $1 billion dollars. These statistics make it apparent that domestic violence is quietly draining the bank accounts of corporate America.

What You Can Do About It

What are security directors and business leaders doing about it? Fortunately, there are solutions. Many proactive companies and government agencies are addressing their financial loss through employee education. They are training their staff to recognize and appropriately respond to domestic violence. Managers and supervisors are taught to implement policies and procedures that address domestic violence as a workplace issue.

Comprehensive training can help you minimize your liability and assist your employees to become more productive and healthy. If violence should erupt, you will be able to demonstrate that your company acted in good faith by providing your employees with:

- The training necessary to recognize and address domestic violence
- Channels through which injured employees or concerned co-workers can seek safety and assistance

As a tool, conducting employee training is one of the least expensive forms of insurance available to your business. Through proper training and policy development and implementation business owners can create a work environment that supports battered workers, while alleviating the financial impact of domestic violence in the workplace. The end result will be a healthier, more productive work force and a substantially reduced legal exposure for your company. (Learn more about domestic violence at www.aidv-usa.com)

Coleen & Rick's Five TIPS to Protect Employees from Domestic Violence

1. Establish an environment of support for your battered workers. They need to understand that your company will not tolerate domestic violence. Display posters, brochures, and safety cards in restrooms or lunchrooms so battered workers can access information without having to ask for it.
2. Ensure your Employee Assistance Program (EAP) has victim-friendly policies. The HR staff must receive enhanced training on how to deal with battered women. Develop a list of local resources for victims and perpetrators.
3. Train your employees and management-level staff to identify domestic violence in the workplace. Teach them how to appropriately assist a battered co-worker.
4. Make sure all supervisors and managers receive special training to ensure compliance with numerous federal and state regulations regarding domestic violence victims and the workplace.
5. Implement a "Domestic Violence in the Workplace" policy or supplement the current workplace violence policy. Contact the American Institute on Domestic Violence (AIDV) at the email address below for a free sample policy.

The American Institute on Domestic Violence helps employers address domestic violence in the workplace via on-site employee and supervisory training. Coleen Widell & Rick Naylor are available to consult with your human resource or security staff to develop a practical and affordable response to this workplace threat.

Crisis Intervention
By Scott Castleman

When dealing with the criminal element, in any capacity, whether you are a business owner, police officer, private investigator, or a private citizen, you must be cognoscente of unforeseen dangers. One can never estimate the way a person will react when being confronted with dishonesty. There are ways, however, to reduce the risk of violent confrontations. In my career as a private investigator, I have been responsible for arresting hundreds of criminals for a variety of different offenses, from shoplifting to auto theft and everything in between. Unfortunately, we do not have the visible presence of a police officer in uniform, which in most instances has the tendency to de-escalate the situation. As citizens, we must rely on a variety of techniques in an effort to maintain control over a possibly explosive situation. Out of the hundreds of criminals I have arrested, not one reacted the same way to being detained. Unfortunately, it is this element of surprise, which makes detention of criminals difficult. We must maintain a proper balance of the low-key approach, but always be ready to defend ourselves if necessary. We do not want to take a tactical approach to detention on every occasion, which would ultimately make us look silly and appear unprofessional. It is finding this happy medium, which becomes a difficult task to anyone facing the dilemma of placing someone into custody.

Interpersonal communication skills are essential to an investigator in order to communicate effectively with emotional, angry, or hostile shoplifters. These skills will also assist in de-escalating a crisis situation.

There are *five phases of tactical communication*, which should be utilized when dealing with difficult arrest situations.

Identification: Let the suspect know who you are and why you are speaking with them. If they know that you actually witnessed their theft, they are more likely to cooperate.

Persuasion: Attempt to motivate the subject into acting or reacting a certain way. People can be motivated and persuaded to act in a certain manner.

Encouragement: Ask them to accompany you to the office before ordering them. Some people require encouragement before they will take action.

Enforcement Action (verbal or physical): *Verbal* enforcement action can be limited to demanding the subject accompany you to the security office. *Physical* enforcement will consist of placing the subject in handcuffs through force.

Conclusion: The final step is as important as all the prior steps. In an effort to ensure a positive conclusion, even if you were required to take physical action in order to detain the subject, thank the citizen for their cooperation. And escort them from the store.

Keeping in mind the five tactical communication phases, you will be able effectively to detain shoplifters. You must, however, maintain a passive-defensive demeanor. This is difficult to maintain, but it is critical to be effective. Unfortunately, you can never determine whether you will be dealing with a hostile individual or not. Sometimes there will be confrontation regardless of what steps you take to avoid it.

Following is a List of De-escalation Techniques to Follow if the Suspect Becomes Aggressive:

- Intermittent eye contact: This is less threatening.
- Maintain a calm demeanor: Often the suspect will mirror your demeanor.
- Allow the suspect to vent: Only if this form of venting is non-violent and non-abusive. This may reduce the potential for violence.
- Listen attentively: Many criminals will try to explain why they committed the theft—"it was an accident" or whatever. Listen to what they have to say and appear concerned.
- Allow personal space: Respect the subject's personal space. They may become less agitated.
- Non-threatening posture: Maintain a relaxed detention and interview stance.
- Reinforce cooperative behavior: They will repeat behavior for more praise or reinforcement.
- Don't tell the person to "calm down": This will only escalate the hostility.
- Avoid using "I", "me", or "you": It is less aggressive and less challenging.
- Identify a common enemy or problem: Demonstrate you are not the enemy.
- Give the person a way out or a way to shift the blame: Reduce their need to defend their ego.
- Isolation: If there is no audience, there is less need to "grandstand."

- Set conversational limits: Do not allow threatening or abusive speech.
- Be an active listener.
- Be aware of non-verbal threats: Look for tightened fists or clenched jaws.
- Be consistent.
- Use distractions: Determine what piques their interests.

Once you make the initial contact it is essential that you show them proper identification or badge (if applicable) and advise them of your position i.e. store manager, private investigator, loss prevention etc. If dealing with a shoplifter, for example, you would want to confront them in a non-accusatory manner by asking, "Did you forget to pay for something today?" By phrasing the question in a non-accusatory manner you allow them an excuse for their theft. "Yeah, you are right; I forgot to pay for my smokes." Although you know full well that they had no intentions of paying for the item, you should phrase your statement as such. The next step is to escort them to the security office. Always walk behind them and give them verbal instructions as to where they are going. Never walk in front of them, this is a serious safety hazard and allows them a perfect opportunity to "ditch" additional merchandise, drugs, or weapons. Try to maintain a conversation with them to keep their attention focused. This is generally when the shoplifter is making a decision if he is going to attempt to flee or fight with you in an effort to get away. By carrying on a conversation with the shoplifter you are diverting their attention.

Now, you must consider what your course of action will be if your suspect becomes aggressive or non-cooperative. There is no real standard way to deal with aggressive or physical confrontations. Obviously, make sure you are familiar with any company policies or state laws they apply to the Use of Force. For the most part, according to most state laws relating to citizen's arrests, you may use justifiable and reasonable force in order to detain a criminal. Does that mean if they strike at you, you may strike back? Absolutely. However, I do not recommend getting into physical altercations with the criminals. Unfortunately, it is sometimes necessary in order to protect yourself, or your partner's safety. Getting your ribs broken or your teeth knocked out or even getting killed in order to recover a $3.00 pack of cigarettes is ridiculous. Unfortunately, there are some circumstances where physical restraint is necessary. Always use caution when deploying the use of physical force and only use the force necessary in order to restrain or detain the individual. Mace is an effective restraint. Just be sure that you are aware of the laws in your state as it pertains to pepper defense spray.

The key to an effective detention is to remain as low key as possible when making the stop. Always be firm, but try not to attract unwanted attention from customers or pedestrians. Remember that innocent bystanders may be in harm's way if your suspect decides not to cooperate.

Any arrest has the potential to escalate quickly into a physical confrontation. As discussed earlier, remaining in a calm, yet confident demeanor, will drastically decrease the chances of the situation escalating into a physical confrontation.

Ordinarily a passive situation goes awry based on a statement or action that the investigator makes. The following is a list of investigator's actions that may cause an escalation of violence during an arrest.

- Prolonged eye contact: An aggressive gesture will most likely cause the criminal to meet the challenge.
- Shouting at the subject: If you shout at the subject, they will shout back. Since you are the professional, you will appear as the foolish one, not only to the criminal, but to bystanders as well.
- Sarcasm: When sarcasm is used on you, what is your reaction? Are you more likely to cooperate?
- Making threats: The moment they call your bluff, you have instantly lost any credibility.
- The use of profanity: This is a sign that you are out of control or fearful. Both reactions are almost impossible ways to regain control.
- Overreacting: To overreact is to lose credibility as the person in charge.
- Indifference: If they sense that you do not care, real or imagined, they will become angry or hostile.
- Offensive contact: This form of unjustified contact sets you up for a lawsuit.
- Aggressive body language: If you intimidate the suspect, human nature is to fight or flee.
- Ego contests: This only demonstrates an immature and unprofessional attitude. This is also known as a "PI'ssing match."

Once you have successfully escorted the suspect to a holding area, a pat down search should immediately be conducted of the suspect (when applicable). It is important to understand that once a private investigator or employee of a business acts under the guidance or direction of a law enforcement officer, the investigator or employee becomes an agent of that law enforcement officer and must comply with all search limitations of law enforcement. Further, if the investigator or employee's actions are unlawful, the courts can suppress the evidence or statements obtained. Unfortunately, not only should you be aware of the laws, you must also make yourself familiar with company policies. For example, in an effort to cover you from civil liability, if you are a male and have a female in custody, a female employee should *always* be present and vice versa.

Prior to conducting a search of the suspect, you must obtain a "consent to search." This may consist of verbal or written consent. Prior to conducting your search you must verbally ask the suspect if they are in possession of any weapons, needles, or sharp objects before you search them. Often they will reply "yes" and show you exactly where the needle is located. Obviously, at this point you should wear gloves when recovering the needles. Sometimes, however, they will advise you that they do not have needles, and you will find them anyway. Usually their excuse will be "I did not know they were there, this is not my jacket." If this does occur and they are not currently in handcuffs, now would be a good time to place them in wrist restraints for their dishonesty. Who knows what else you will find?

If you encounter weapons, the suspect should also immediately be placed in handcuffs. It will always be your decision when to utilize your wrist restraints. Many investigators place shoplifters in handcuffs every single time for safety. I, on the other hand, only use them when I feel threatened. However, what threatens me may not threaten you and vice versa. Only you will be able to determine when wrist restraints will be used.

Once you have completed a thorough pat down of your subject, you should begin your questioning. Although you will be interrogating the subject, it should never appear as an interrogation. Interviewing and interrogating suspects is a refined art form. Not all investigators will have the skills to be effective interrogators. Investigators should also be familiar with the differences between interrogations and interviews.

Interrogations vs. Interviews
An *interview* can be described as a form of asking questions in a non-confrontational manner.

An *interrogation* can be described as a process of asking questions in a more confrontational manner. This is a far more aggressive approach to the passive interview technique.

Often an investigator will conduct a combination of the both. He or she will initially start out with an interview and then begin a more aggressive interrogation. This is generally done during the interview when the suspect displays signs of confessing.

When a citizen feels that he or she is compelled to engage in conduct from which he or she has a legal right to abstain from, the court may construe the investigators' actions as *coercion*.

To become a successful interviewer it can take years of practice to refine the art. There are two main functions of the interview when it pertains to shoplifters. The first is to get a confession, and the second is to prove intent. Generally, intent can be proven if the suspect has no money on their person. Obviously, this will mean that he had the intent to steal upon entering the store, knowing full well he had no money. Unfortunately, it is not that easy. A jury wants to hear evidence of an admitted form of intent. Obviously, asking the suspect if he intended to steal the merchandise is not the wisest idea. Apprehending shoplifters is only a small part of the investigator's job. The most difficult part is securing a confession. A good interrogation can reveal a wealth of information. Shoplifters are generally not stupid and have been through the court system before. However, with a little persistence, you may be able to obtain information regarding other boosters that are hitting your stores or a list of markets or fences that are buying shoplifted merchandise.

Most investigators come across as naive when they interview subjects. The shoplifters have a keen sense of whether they are dealing with a rookie or an experienced investigator. If they think they are dealing with a rookie, they will either not talk or give you misinformation.

There are four main rules to follow when interviewing a suspect:

- Always maintain control of the interrogation, and of yourself.
- Have confidence in yourself.
- Always be truthful with the suspect; never make promises you cannot keep.
- Maintain proper perspective.

Identifying Dishonesty

Probably the most difficult aspect of an interview or interrogation, is determining if the suspect is lying to you. Unfortunately, we live in a deceptive society, in which there is no real punishment for lying. If a shoplifter is caught in a lie, generally the reports must be re-written and no added punishment is placed on the shoplifter for their dishonesty. So, look at the odds. They have about a fifty percent chance that you may not be able to uncover the truth or their true identity in the short time frame in which you have them in custody. Approximately 60% of the shoplifters I detain have lied to me about one thing or another. It may be a simple lie, such as "No, I do not have any needles in my pocket." "No, I have no warrants for my arrest." or "No, I have never been arrested before." On the other hand, sometimes the lies are far more substantial, such as giving the arresting investigator a fake name and fake identification to match the fake name. Obviously, this takes more skill to detect. So, how do you know if you are being lied to? There is no easy answer to this question. However, there are a few signs and indicators to look for to determine dishonesty.

Indicators of dishonesty

- Little or no eye contact
- Limited physical expression i.e., few head, arm, and hand movements. Physical expression will seem stiff, almost mechanical. Also, the hands and arms are pulled in towards the body.
- Shrugging or nodding responses instead of verbal. Also, the suspect may do a combination of both, but the gestures will probably not match the verbal response.
- Excessive yawning (which appears to be choreographed).
- The timing may be off between gestures and words.
- Eyes move away from the accuser, generally towards an exit.
- Generally, the suspect will be reluctant to face his accuser and will generally turn his head or shift his body.
- The suspect will use your words to make his point.
- Excessive pauses before answering a question. Dishonest responses take longer to make up.
- Use of phrases, such as "To be honest with you." "To tell the truth", etc.
- The use of humor or sarcasm to deflect your questions.
- When a subject or question is changed, they are in a better, more relaxed mood.
- Excessive stalling and asking you to repeat the question several times. Or, saying "I don't understand the question."
- Excessive fidgeting.
- Playing with their hair or rubbing their face or neck.
- Nervously tapping their feet.
- Wiping the sweat from their palms onto their pants.
- Repeating the word "what" after every question you ask, as if they did not hear your question. This is a common stalling

method in order to think up a quick lie to your question. This method is common to juveniles.

If you have two shoplifters working together in custody at one time, immediately separate the two suspects. This is known as "divide and conquer." Gather all of the information about suspect #1—name, address, etc. Then interview suspect #2 and ask them the same questions regarding suspect #1. More often than not, suspect #2 will say that he or she does not know the other suspect very well and just "hitched" a ride with them. Prepare for this response, since it is common.

In order to be effective in interviewing suspects and gaining their trust you must begin by building a rapport with suspects. Often this cannot be achieved, since no matter how hard you try to develop a rapport with them, they will not comply. Rapport creates trust, which allows you to build a psychological bridge to the suspect. Always keep in mind that when dealing with professional shoplifters, your rapport with one suspect will be passed on to all other boosters. If you disrespect them or go back on a promise and break your rapport, you will be forever labeled as a liar. The next booster you have in custody will know you. Trust me on this. Create a rapport and stick with it.

If you know that you cannot work a deal with the suspect, do not even bring it up. There is no honor among thieves. There should be honor among investigators and criminals. This is the only tool you will ever have to gain confidence and gather information that is not ordinarily divulged by the criminals. In cases where this is possible, you must follow a few guidelines in order to establish your rapport.

The way you present yourself is crucial to establishing your rapport. Simple things, such as unbuttoning your coat or uncrossing your arms can make a person feel less defensive.

Matching posture movements. Attempt to match the suspect's rate of speech. If he is conversing in a slow, relaxed tone, you should do the same. If he is speaking rapidly, you should attempt to match his speech.

Matching key words. If your suspect is prone to using specific words or phrases during the course of a conversation, employ them when you speak. For example, if the suspect says "You know what I'm saying?" frequently in conversation, you should also use the phrase when asking them questions. If they are prone to using profanity in their speech, use profanity to explain your point.

Once you have established your rapport and have begun your questioning, you may begin your report writing.

Keep in mind that if you ever are placed in the situation of arresting a suspect, you must try to obtain some sort of formal training in interviewing, or hire an experienced private investigation firm that is used to dealing with such cases. The fact is that a little mistake can be financially devastating by way of lawsuits on behalf of the suspects. Know what you are doing or retain someone that does to handle the criminal issues.

Scott R. Castleman is the owner of the Corporate Crime Control Association in Oregon State, specializing in employee theft and protection. He has authored two investigative publications available at Thomas Publications or www.thomaspub.com. Castleman has been an investigation instructor at Oregon's largest Private Investigation Academy for over six years. He has received letters of appreciation from the City of Portland and from the Commander of the Portland Police for his efforts in fighting crime on the streets of Portland.

Preventing Sexual Harassment Lawsuits

By Barbara Kate Repa

Public hand wringing has quieted some since the issue of sexual harassment on the job catapulted into the forefront of public consciousness in the early 1990s. The cataclysmic event was Professor Anita Hill's charge that Judge Clarence Thomas, then a nominee to the U.S. Supreme Court, had sexually harassed her when the two worked together at the Equal Employment Opportunity Commission (EEOC)—ironically the agency charged with investigating and curbing sexual harassment on the job.

But the debate rages on in America's workplaces. Ask employers what potential problem haunts their dreams these days, and most will offer the same unequivocal response: sexual harassment.

There is good cause for unrest. The number of claims filed with the EEOC has soared since 1991—the year when the Thomas-Hill hearings made sexual harassment a workplace buzz phrase and the Civil Rights Act was amended to allow damages awards to those who make out a case.

The increase in the number of claims filed does not necessarily mean that more harassment is occurring. But it undoubtedly signifies increased awareness of the issue and a greater willingness to take action against it. This reality should act as a wake-up call to all employers: Be Prepared. Those who choose to ignore the problem in the hope that it will go away will find it only gets worse.

This article sets out the policies and procedures an employer can adopt to monitor, detect, and deal with sexual harassment on the job. And the message is hopeful. Conscientious employers can do much to avoid the volatile problems of workplace sexual harassment by taking proactive steps. Crafting a solid policy defining and prohibiting the behavior and scrupulously enforcing it are essential and effective in stamping out the misconduct.

Sexual Harassment Defined

In legal terms, sexual harassment is any unwelcome sexual advance or conduct on the job that creates an intimidating, hostile, or offensive working environment. Simply put, sexual harassment is abusive conduct related to an employee's gender that a reasonable man or woman should not have to endure. And unwantedness is key: behavior that is encouraged or solicited is rarely found to be true harassment.

The forms of behavior that sexual harassment can take are as varied as a perverse imagination can create. An employee whose boss threatens to withhold a promotion unless granted a sexual favor has been harassed, as has one whose co-workers regularly tell offensive, sex-related jokes after being informed they're offensive. An employee who has been pinched or fondled against his or her will by a co-worker has been harassed, as has one whose colleagues regularly leer or threaten to block passageways.

An employee who is constantly belittled and referred to by sexist or demeaning names has been sexually harassed, as has one who is subjected to repeated lewd or sexually explicit remarks. Sexual harassment occurs when a supervisor acts as if the men working under him owe him sexual favors, and it also occurs when a co-worker attacks or intimates a woman because he doesn't think she should be doing what he considers "a man's job."

Although it is usually men who sexually harass women on the job, such abuse can also be inflicted by women against men, by women against women, and by men against men. Hostility based on gender is the test, rather than the gender of those involved.

While a relatively small proportion of cases have proceeded all the way through to a court hearing, those that have reveal a wide variety of harassing activity, covering all genders and walks of life. Those claiming they were harassed include:

- a female employee in a Florida shipyard who complained that the workplace walls displayed pictures of a woman's pubic area with a meat spatula imposed over it
- a female secretary whose boss—a lawyer—repeatedly bullied and threatened her and grabbed her as he filled her breast pocket with M&Ms
- a male roustabout onboard an oil rig who was sexually assaulted by several co-workers as he showered
- a number of women who were subjected to a dress code specifically established by their supervisor so that he could admire their legs, and
- two male employees whose supervisor threatened to fire them if they refused to participate in various antics including strip poker at the work site after hours.

The Employer's Duty

In a number of recent pronouncements on sexual harassment, the U.S. Supreme Court has begun to define what steps concerned employers must take to help stamp out this unwanted behavior in their workplaces.

In one case the Court held that an employer may be liable for sexual harassment even when an employee did not succumb to sexual advances or suffer adverse job consequences. But the employer could defend itself against liability and damages by showing that it used reasonable care in stopping harassment—a strong written anti-harassment policy and investigation procedure. And the Court leveled the playing field, opining that an employee who does not take advantage of an existing workplace policy by reporting the harassment has a considerably weaker case. *Burlington Industries, Inc. v. Ellerth*, 524 U.S. 742 (1998)

In another case the Court held again that an employer could defend itself against a sexual harassment charge by showing it acted reasonably to prevent it. But it found that an employer acted unreasonably by both failing to distribute its anti-harassment policy and by not establishing a complaint procedure—and, as a result, was liable for the harassment. *Faragher v. City of Boca Raton*, 524 U.S. 775 (1998)

And in a third case the Court held that scrupulous employers can save money too, ruling that employers who have a strong written policy against harassment, in addition to a current training program for employees, can shield themselves from the harsh potential liability of punitive damages. *Kolstad v. American Dental Association*, 527 U.S. 526 (1999)

Employers anxious to protect themselves can find specific guidance in another recent case, this one decided by a California appellate court in which an employer was commended for its "textbook example of how to respond appropriately to an employee's harassment complaint." The court was impressed that immediately after receiving a letter from an employee alleging she was sexually harassed, the employer:

- requested a meeting with her to discuss the allegations
- questioned the accused harasser
- interviewed employees and former employees the accused harasser supervised, and
- promptly gave the accusing employee a letter summarizing its investigation and the action it took.

While the company found no clear evidence of sexual harassment, it did find the accused had exhibited extremely poor judgment in a number of words and deeds. It added a written reprimand to his personnel file specifying the errors he made and warning him from initiating contact with the accuser—on pain of immediate termination. *Casenas v. Fujisaw USA, Inc.*, 58 C.A. 4th 101 (1997)

Sexual Harassment Policies

Courts will occasionally require a company to write up a comprehensive policy as part of the legal relief ordered in a sexual harassment case. But more often these days the impetus to rid the workplace of sexual harassment—or to prevent it from occurring—comes from within. A well-crafted sexual harassment policy can offer both employees and employers guidance and certainty in dealing with harassment in the workplace.

Benefits to Employers

Employees who are sexually harassed on the job often become distressed, depressed, frightened, and angry—or some combination of these. It is not easy to hold a job while also working full time to keep a harasser at bay. And the challenge often takes its toll. Sexually harassed workers often become demoralized and miss work, and even while on the job they are much less able to concentrate and to work efficiently.

All of this runs counter to nearly every employer's prime goal: workplace productivity. A number of recent studies of women who claim they were harassed on the job show the high cost to employers of tolerating it:

- Over 25% of women who are sexually harassed use leave time to avoid the uncomfortable work situation.
- At least 15% of women who are harassed in the workplace quit their jobs because of it.
- Nearly 50% of those harassed try to ignore it—and end up suffering about a 10% drop in productivity.
- The harassed woman's peers who know of the situation also suffer a 2% drop in productivity.

But perhaps money talks most loudly and chillingly: Ignoring sexual harassment can cost the average company up to $6.7 million a year in absenteeism, employee turnover, low morale, and low productivity, according to recent workplace studies.

In addition to these indirect costs an employer who tolerates sexual harassment risks the high administrative costs involved in EEOC and other agency complaint investigations, as well as the pricey possibility of a successful lawsuit against the company. Harder to quantify—but perhaps most damaging—can be the sullying of the company name and losing goodwill if the charges go public.

Since many employers hire expensive lawyers at the first scent of legal trouble, the costs of defending a sexual harassment lawsuit are extremely high. And of course, if the harassed employee wins, costs to the employer will be even higher—sometimes sky higher. Several employers have been hard hit with jury verdicts ranging into several million dollars for emotional distress, lost wages, and wrongful discharge.

Benefits To Employees

Despite some paranoid perceptions, very few workers who bring complaints about sexual harassment on the job are seeking big bucks or vengeance. Their goal, at least initially, is to make the harassment stop—and to take all-practicable steps to see that it stays stopped. It is typically only when an employer offers no fair way to resolve the problem—or at least get it fully aired that sexual harassment complaints evolve into lawsuits.

If an employer has a clear written policy prohibiting harassment, all who work there will have a firmer idea of what behavior will not be tolerated on the job.

Tips For Good Policies

An effective sexual harassment policy discourages harassment—pure and simple. And it should also encourage employees to promptly report all serious incidents of harassment. Employees cannot be required to report sexual harassment; that would be coercive. If you have a sensitive and sensibly written policy against harassment, backed by good policies to deal with it as confidentially as possible and to prevent retaliation against the person making the complaint, most employees will gratefully comply.

A model policy is offered here, but keep in mind that a sexual harassment policy will be effective only if it is appropriate and realistic for your workplace. A conservative, multi-tiered accounting conglomerate based in New York City, for example, may find it reasonable to have more buttoned-down strictures on permissible workplace behavior than a small computer graphics company in San Francisco or a tractor dealership in Pocatello, Idaho.

Also, to be effective, a sexual harassment policy must be clear and comprehensible. A growing number of employers pridefully point to the fact that they have policies prohibiting sexual harassment in place. However, a quick perusal of most of them revels a sad truth: They are so replete with legalese that they confuse more than they cure. Be sure your company adopts a policy that is meaningful—and written so that it can be easily understood by everyone. It is a good idea to ask a cross-section of employees to read a policy for clarity before you distribute it to the workplace in general.

Finally, all employees should be encouraged to view the sexual harassment policy as a work in progress. It will surely need to be changed with the times and perhaps altered to reflect the changing roles and numbers of employees. To make sure that this happens, be sure to review the policy annually and amend it if need be.

Elements Of A Good Policy
The most effective sexual harassment policies have a number of common elements.

Zero tolerance
The policy should include a statement that sexual harassment is discrimination, illegal, and will not be tolerated. A short, direct statement is most effective.

Clear definitions
It is essential to define what behavior is prohibited. This used to be no small task. Fortunately, however, after struggling for years to come up with meaningful definitions of prohibited harassing behavior, the EEOC, legislatures, and courts have made considerable progress. Today it is possible to provide plain speak. While even the best policy cannot set out every kind of prohibited behavior, a good one should go beyond declarative language to list some specific examples. For example, a sexual harassment policy may ban:

- Verbal harassment, including making sexual comments about a person's body, telling sexual jokes or stories, spreading rumors about a co-worker's sex life, asking or telling about sexual fantasies, preferences, or history
- Nonverbal harassment, such as giving unwanted personal gifts, following a person, staring at a person's body, and displaying suggestive material such as pornographic photos
- Physical harassment, including touching yourself in a sexual manner in front of another person or brushing up against another person suggestively.

Specify consequences
Spell out what action may be taken for first offenses and for repeated unacceptable conduct. Punishments should be appropriate,

certain, and reasonably strict. Appropriate action can range from a written warning in the employee's personnel file to counseling, suspension from work, transfer to a different shift or position, or dismissal.

Reporting instructions
Give guidelines on how to report harassment. Detail how, when, and where employees can complain about behavior that makes them feel uncomfortable. For example, if the first step is to contact a particular person within the company, make clear how to do that. The complaint process should be as confidential as possible. If you want employees to use a form to initiate a complaint, include a copy in every employee handbook—or at least be sure that the form is easily accessible.

Provide an alternative to employees who do not wish to file a complaint with their own supervisors—some of whom may either be responsible for the harassment or guilty of ignoring it. Make it possible to report the harassment to any other supervisor—or perhaps to anyone on a panel of employees designated to handle company grievances.

Investigate promptly
Provide for prompt and confidential investigations. Despite all best intentions to keep the matter hushed, news of a harassment investigation tends to circulate fast. And people often take sides quickly—which can be deadly to workplace morale. Proceeding speedily and appropriately with the investigation helps take the strain from the workplace, stems the tides of worry and gossip, and, above all, is the best way to get at the truth of the matter. During the investigation, as during the initial complaint filing, make every effort to respect the confidentiality of all involved.

Make the results known
Give both the complaining employee and the accused harasser a brief report of the findings of the harassment investigation and a notation about whether any action is to be taken.

Protect against retaliation
Take a strong stand against retaliation. It's the fear of being ostracized or punished that keeps many employees from reporting even the most egregious workplace harassment.

A good harassment policy should assure employees that there would not be any retaliation for filing a complaint against workplace harassment or for cooperating in the investigation of another's complaint. The policy should also spell out the kind of discipline—reprimand, suspension, transfer, and dismissal—that may be imposed against those who retaliate.

And beware that retaliation may come from unexpected people and places. It may come from the harasser who, as a manager or supervisor, transfers an employee to an undesirable location, changes his or her work duties to be either mundane or overwhelming, or fires the worker.

Co-workers—especially friends and supporters of the alleged harasser—sometimes retaliate against an employee who has rocked the boat by reporting the bad behavior in the first place. They may refuse to cooperate with the employee so that it becomes ever more difficult for him or her to work, or they may shun the worker who complains completely.

Here is a good start for a sexual harassment policy that you can modify to meet the needs of your workplace.

Sexual Harassment Policy

[Employer name] is committed to providing a work environment in which all employees can work together comfortably and productively, free from sexual harassment. Such behavior is illegal—and will not be tolerated here.

This policy applies to all phases of employment—including recruiting, testing, hiring, upgrading, promoting or demoting, transferring, laying off, terminating, paying, giving benefits, selecting for training, travel or social events.

Prohibited Behavior

Prohibited sexual harassment includes unsolicited and unwelcome contact that has sexual overtones. This includes:

- written contact, such as sexually suggestive or obscene letters, notes, e-mail messages, or invitations
- verbal contact, such as sexually aggressive or obscene comments, threats, slurs, epithets, jokes about gender-specific traits, sexual propositions
- physical contact, such as intentional touching, pinching, brushing against another's body, impeding or blocking movement, assault, and coercing sexual intercourse
- visual contact, such as leering or staring at another's body, gesturing, displaying sexually suggestive objects or pictures, cartoons, poster, or magazines.

It is impermissible to suggest, threaten, or imply that failure to accept a request for a date or sexual intimacy will affect an employee's job prospects. For example, you should not either imply or actually withhold support for an appointment, promotion, or change of assignment, or suggest that a poor performance review will be given because an employee has declined a personal proposition.

Also, you may not offer benefits, such as promotions, favorable performance reviews, favorable assigned duties or shifts, recommendations or reclassifications, in exchange for sexual favors.

Harassment By Others
In addition, [Employer name] will take all reasonable steps to prevent or eliminate sexual harassment by non-employees, including customers, clients, and suppliers who are likely to have contact with our employees on the job.

Monitoring
[Employer name] will take all reasonable steps to see that this policy prohibiting sexual harassment is followed by all employees, supervisors, and others who have contact with our employees. This prevention plan will include training sessions, ongoing monitoring of the work site, and a confidential employee survey to be conducted and evaluated annually.

Discipline
An employee found to have violated this policy will be subject to appropriate disciplinary action, including warnings, reprimand, suspension or discharge, according to the findings of the complaint investigation.

If an investigation reveals that sexual harassment has occurred, the harasser may also be held legally liable for his or her actions under state or federal anti-discrimination laws or in separate legal actions.

Retaliation
Any employee bringing a sexual harassment complaint or assisting in investigating such a complaint will not be adversely affected in terms and conditions of employment and will not be discriminated against or discharged because of the complaint. All complaints of retaliation will be promptly investigated and punished if deemed necessary.

Complaint Procedure and Investigation

[Title of person appointed] is designated as the Sexual Harassment Counselor. All complaints of sexual harassment and retaliation for reporting or participating in an investigation shall be directed to the Sexual Harassment Counselor or to a supervisor of your choice, either in writing, by filling out a Complaint Form, or by requesting an individual interview. All complaints will be handled as confidentially as possible. The Sexual Harassment Counselor will promptly investigate and resolve complaints involving violations of this policy and recommend to management the appropriate sanctions to be imposed against violators.

Training

[Employer name] will establish yearly training sessions for all employees concerning their rights to be free from sexual harassment and the legal options available if they are harassed. In addition, training sessions will be held for supervisors and managers, educating them in how to keep the workplace as free from harassment as possible and in how to handle sexual harassment complaints.

Access to Policy

A copy of the policy will be distributed to all employees and posted in areas where all employees will have the opportunity to freely review it. [Employer name] welcomes your suggestions for improvements to this policy.

Complaint Procedures

A good harassment policy is a crucial first step in ridding a workplace of offensive, unwanted behavior. But it won't count for much unless you also adopt a trustworthy and energetic procedure for handling and investigating complaints.

A helpful attachment to a sexual harassment policy is a Complaint Form to record important dates, facts, and names and flag important follow-up procedures.

Sexual Harassment Complaint Form

Name: _____

Department: _____

Job   Title:

Immediate Supervisor:

___

Who was responsible for the harassment?

___

Describe the first incident:

___
___
___
___

Approximate date, time, place:

___

Any other people present:

___

What was your reaction?

___
___
___

Describe the second incident:

___
___
___
___

Approximate date, time, place:

___

Any other people present:

___

What was your reaction?

___
___
___

Subsequent incidents:

___

_____
_____
_____

Approximate date, time, place:
_____

Any other people present:
_____

What was your reaction?
_____
_____
_____
_____
_____

I understand that these incidents will be investigated, but this form will be kept confidential to the highest degree possible.
Employee's signature:
_____

Date: _____

For Employer's Use

Dates of investigation of complaint:_____

Date of final report:_____

Date copy sent to employee:_____

Action taken: _____

Date of follow-up conference with employee: _____

Results:_____
_____
_____
_____

Date of follow-up conference with employee: _____

Results:_____
_____

Date of follow-up conference with employee: _____

Results:_____
_____
_____
_____

Sexual Harassment Training

Sexual harassment in the workplace can be curbed or even eliminated by training supervisors and all other employees to recognize its signals and head it off. Doing this can help curb or eliminate harassment-related lawsuits, increase productivity, and make the workplace more comfortable for everyone.

Sexual harassment training can take many different forms, depending on the size and budget of your company. The best training programs involve all levels of employees, so that the entire workplace gets a better understanding of the issue—what it is, why it happens, who it affects, typical reactions of both the harasser and harassed, and effective ways to make the destructive behavior stop.

An increasing number of employers now offer anti-harassment training on the job—presented by individual trainers onsite, delivered to groups in seminars, and increasingly through an Internet-based training product. While most employers with moxie now know that providing employees with targeted training is one clear way to limit their liability for workplace discrimination and harassment, it is far less clear how effective training should look and feel. Here is a summary of the elements—some obvious, some less so—that courts have recently decreed are required for this compliance training to be sound and effective.

Once Is Not Enough

The faces in most workforces change these days, even more rapidly than in the past. So unless employers make special efforts to train new hires about how to recognize and report discrimination and harassment, for example, those new workers will likely fall through the cracks of awareness and remain untrained. And their employers will be left dangerously vulnerable to liability.

Also, because state and federal laws controlling prohibited workplace behavior change frequently and are often convoluted and difficult to remember, according to workplace experts it's also important to repeat training on poignant workplace issues such as discrimination at least yearly.

Employers who fail to offer repeated training risk being unable to establish that they made essential good faith efforts to clamp down on the illegal behavior on the job. For example, a Texas court recently disregarded Wal-Mart's claim that it provided adequate harassment training after a manager there hazily testified to attending only one brief course on it "a couple years ago" and the alleged harasser himself recalled only a scant 15 minutes of training on the company's anti-harassment policy. *Wal-Mart Stores v. Davis*, 979 S.W. 2d 30 (1998)

Make Sure the Trainers Are Trained
Training that is boring, glib, or overly simplistic is clearly lost on those trainees who spend more time rolling their eyes than digesting the finer points.

But when the substance is unclear or downright wrong, the training delivered is legally tantamount to no training at all. For example, a Kansas court recently held that a local home improvement company could not demonstrate its good faith efforts to train its employees about the rights and wrongs of sexual harassment. The court reached its decision after the company's training manager testified that she believed a male supervisor would not commit sexual harassment if he either exposed his genitalia to a female subordinate or grabbed her breasts, as long as he apologized afterward. *Cadena v. The Pacesetter Corporation*, 224 F. 3d 1203 (2000)

Share And Share Alike
Because courts assume, for better and for worse, that supervisors represent the essence of a company, employers must be especially certain that they are adequately trained in what constitutes illegal conduct on the job. The potentially costly legal impetus for this is that companies can be held vicariously liable for a supervisor's harassing or discriminatory behavior—whether or not anyone else in company management was aware it occurred.

In reiterating this view recently, the U.S. Supreme Court held that the city of Boca Raton, Florida could have protected itself from liability in a sexual harassment case if it could have demonstrated that its male supervisors, who had repeatedly fondled and hectored their female subordinates, had been given adequate training in the wrongful behavior. *Faragher v. City of Boca Raton*, 524 U.S. 775 (1998)

But some companies, often as a cost savings measure, provide only managers with training, leaving out the rank and file. This can often misfire as a pound-foolish mistake, since non-managers file the grandest majority of claims. And they can easily win courts' sympathy—and subject employers to liability—if they can show that the very souls accused of harassing them were never trained to recognize that the behavior is illegal.

Make It Make Sense

Many well-meaning employers have adopted training programs, but they overlook the obvious: The training must be comprehensible to those being put through its paces.

Courts, arbitrators and agencies recently called upon to evaluate the legitimacy of specific compliance training programs look to reality. And they have found training programs to be wanting, for example, on the grounds that they were:

- so legalistic that they made little sense to the lay audience for which they were intended
- delivered only in English to trainees whose native language was not English
- given orally to trainees with severe hearing impairments

Track the Trained

Even the best training may mean nearly naught from a legal standpoint if an employer is not able to show which employees were trained and when they received the instruction. If the fact or timing of training is ever questioned, an employee who claims to have been overlooked will often seem the most sympathetic. Only companies that can back the fact of training with ironclad tracking systems are likely to prevail.

Monitoring The Workplace

In addition to adopting a sexual harassment policy and complaint procedure and providing employees with training, employers who are truly committed to eradicating sexual harassment must actively monitor their workplaces.

The best way to do this is to regularly remind employees that sexual harassment is illegal and will not be tolerated by redistributing your sexual harassment policy. You may also consider conducting a survey of the workplace every year or so—encouraging employees to give confidential assessments of whether sexual harassment affects their work lives. And heed their responses: The best changes are often suggested from within.

Attorney Barbara Kate Repa is an employment mediator in San Francisco, California and the author of several books on the workplace, including: *Sexual Harassment on the Job: What It Is and How to Stop It, Avoid Employee Lawsuits, Firing Without Fear,* and *Your Rights in the Workplace.* All are published by Nolo Press in Berkeley, California.

Responding to a Robbery
By Edmund J. Pankau

In recent years robberies of financial institutions and retail establishments have increased dramatically. For the employees of any type of business that has been robbed, no event can produce more stress. Whether or not the robber is armed, whether or not there is a physical threat, the act of robbery produces a set of circumstances that are extremely dangerous and can cause a wide range of side affects that damage employee efficiency and productivity and affect the morale of the business well beyond the occurrence of the actual event.

While there are no known statistics that document the nature and frequency of personnel changes after a robbery, security experts concur that there are frequent requests for transfers and resignations due to crimes of this type, especially if violence occurs. The more serious the robbery, the more likely resignations and transfers will follow.

Based on the examination of many such robberies, security professionals have found that the most successful response to this type of violent crime is post-robbery counseling and employee relations programs that show that the employer understands and cares about the mental health of their employees, enough to provide a program to help them cope with their trauma and overcome their fears. The initial step in such preparation is training to make employees aware of the possibility of robbery and how to handle the situation, should it arise.

Through an understanding of the process of a robbery, and through training by security professionals, employees learn what to expect and how to avoid making rash movements or actions that would put them in further danger or cause an escalation of the violence of the act.

Understanding The Trauma
Anyone who has been through a robbery understands that this act is an extreme violation of their workspace, one that changes their outlook forever. A previously safe work environment suddenly becomes unsafe and can be violated any time—the next minute, the next day, or the next month.

The most often victim of a robbery is going to be the employee out front, as they are the ones that handle the money and the cash registers. They are the first contact for customers and are usually out front, where the action is.

The two most common emotions that go through an employee's mind during a robbery are fear and anger. The level of fear generated through a potentially violent confrontation in a bank, for example, can often bring a form of paranoia that leaves the employee wary of all future customers. They learn to scrutinize everyone who enters the bank and can often become "panicked" when one appears to do something out of the ordinary, once they have become a victim themselves.

Many employees take the robbery as a personal offense and either verbally or physically confront the robber. This type of confrontation is extremely dangerous and often leads to an escalation of the crime.

Once any type of business has been robbed, it is essential to rebuild a sense of security within the workplace. One of the best ways to do this is through post robbery counseling and role-playing reenactment of the crime.

Once a robbery has occurred, employees feel that their world has been shattered. Intense counseling is one of the best means to re-establish a sense of control and bonding that is essential to the rebuilding of employee morale.

One of the best ways to create this rebuilding process is to reenact the robbery at the scene of the crime. By reliving the events of the crime and discussing the steps taken, employees reinforce each other's understanding of the event and bring it into terms of "normalcy" that they can deal with. This process helps employees provide a support process for themselves and allows them to reinforce their belief that they can deal with a potential crisis situation.

Implementing A Crisis Program
There are several ways to implement a post-robbery counseling program and a growing number of resources that are available to the businesses that need such services. Both law enforcement and private security consulting firms have a wide body of knowledge in dealing with these issues and are willing to provide their services to help employees both understand what they went through and to deal with their personal issues brought about by the robbery. In addition, there is a growing trend to deal with this problem internally through the business employees themselves.

Many businesses feel that their purposes are best served by creating internal programs that utilize their own personnel resources because it makes employees feel that management is being responsive to their needs and helps build the bonding process from within the institution rather than from "outsiders."

Since most front office personnel are female, most of them have little or no understanding or knowledge of robberies. By training these employees through a combination of professionally produced video tapes and a "role playing program," supervised by the security staff or outside security consultants, they can generate a greater understanding of the problems and a more positive attitude toward both the crime and the business itself.

The greatest benefit of post robbery counseling is that it helps employees conquer their fears and minimizes the time lost because of nightmares and a general sense of violation.

How can your business, often a small to medium sized office, help employees in dealing with their post crime trauma and fears? Here are a few steps recommended by professionals who have dealt with this problem:

1. Give the victimized employees an opportunity to relocate to another branch or position within the business if it is within a reasonable distance.
2. Allow those employees most affected by the crime a leave of absence, if requested, to deal with their personal issues and fears.
3. Analyze the crime and implement appropriate new security measures. This shows employees the commitment of management and allows a stronger feeling of safety within the institution.
4. Create a series of weekly staff meetings to discuss the crime and its effects on the institution and its employees. Discuss their feelings and their needs to resolve any safety or security issues.
5. Consider the retention of a professional psychologist to attend employee staff meetings, especially in those cases where crimes of violence have occurred or hostages have been taken.
6. If an employee must testify regarding the incident—at deposition or in court—have a supervisor or the owner of the business, someone whom the employee knows and trusts, personally transport the employee to and from the proceedings. This will demonstrate personal support for the

employee and will assist in reducing the additional trauma of being alone in the presence of the perpetrators.

In summary, the best response to a robbery is to help the employees turn the events into a positive experience, through training, counseling, and reinforcement. Show the staff that management cares and is responsive to the needs of both the employees and the business and that it is willing and able to respond to them in their time of need.

Mr. Pankau has been featured in USA Today, Time, Business Week, Wall Street Journal, and New York Times and has appeared on ABC's 20/20, MacNeil-Lehrer, BBC-TV London, CNN-Money Line, Inside Edition, Larry King Live, America's Most Wanted, and many other radio and television programs. A graduate of Florida State University's Department of Criminology, he is a Diplomate of the American Board of Forensic Examiners. He was named by P.I. Magazine as one of the world's top ten investigators.

Workplace Hostage Situations
By Robert K. Spear

Generally people don't like to think about the bad things that can happen to them. Unfortunately, hostage situations in the workplace have become far too commonplace. We can't afford not to think about them. Let's start by addressing the threat. Who are most likely to take hostages in the workplace?

Most Common Hostage Takers

- Disgruntled Employees / Emotionally Disturbed Individuals
- Criminals
- Terrorists
- Ransom Seekers

Disgruntled Employees / Emotionally Disturbed Individuals
This is the most likely threat in many businesses, especially those with a bureaucracy. It has become so prevalent in the US Postal Service, a disgruntled worker who goes off the deep end is said to have gone *postal*. It doesn't matter if the worker has a valid complaint or not—he thinks he does, and that's all that counts. If he chooses to take a boss and fellow workers hostage in order to get what he wants (be they conditions or revenge), he feels totally justified to do so. He believes justice is on his side. Arguing about rights and wrongs isn't a wise thing to do with this kind of hostage taker. Accept his position as valid for the interim and work from that premise.

Emotionally disturbed individuals, especially in domestic disturbances can also be very dangerous. This author remembers the sorority house next to his fraternity. A middle-aged female cook for the sorority looked up, saw her estranged husband walk in with a shotgun, and took off running out of the kitchen. He chased her down on the front sidewalk, blew her away, and ate his own gun. The homicide/suicide was the talk of not just the large state university campus, but was on the evening news statewide. The sorority and the school suffered a fair amount of negative press over the incident.

Often these people have a death wish and want to take as many with them as they can, especially if a boss or fellow worker is perceived to be at the heart of the problem.

Criminals

This is the most common threat to retail businesses such as convenience stores, liquor stores, fast food restaurants, and filling stations. These types of businesses are often targets of opportunity for robbers. If the robbery goes bad and the police show up, the criminal(s) is likely to take a hostage as a way of obtaining a chance for a getaway. If the criminal is a strung-out addict, desperate for a fix, anything can happen.

Terrorists

Traditionally, terrorists have taken hostages to obtain media attention to their cause and to demonstrate how powerless the establishment is. 9/11 and the resulting military actions in Afghanistan and Iraq have changed all of that. Al-Qaida has a new rule of engagement. If an Al-Qaida terrorist comes into your business to take hostages, he won't leave alive and he will kill all his hostages if he can. Publicity is no longer the driving motivator—revenge is.

Ransom Seekers

Your company or organization may be shaken down for ransom money—especially overseas. For profit companies or nonprofit organizations are all at risk. An excellent example of this was the hostage taking of Martin and Gracia Burnham who worked for the New Tribes Mission (a nonprofit organization) in the Philippines since 1986. Martin was a missionary bush pilot and Gracia helped in several different aviation support roles. He was the son of missionaries who worked for the same organization in the Philippines and grew up there. Gracia is the daughter of a Pastor in Arkansas. Martin and Gracia's three children, Zach, 11, Mindy, 12, and Jeff, 15, were all born there.

The Abu Sayyef Group, Muslim rebels, took hostage Martin and Gracia on May 27, 2001. On June 7, 2002, Martin was killed and Gracia was wounded in a sudden firefight between government forces and the rebels. The motives for the hostage taking were strictly monetary. The same may be true for corporations doing business overseas, or it may be to make an environmentalism statement or a political one. As a business, you must be prepared for the possibility and know what to expect.

What to Expect

The following points address what you should expect if you or any of your employees become hostages. We will address the:

- Initial Shock

- Developing Situation
- Stockholm Syndrome

Initial Shock

If you're ever so unfortunate to find yourself in the middle of a hostage situation, you'll need to be aware of the specific emotions and feelings you're liable to have at different times during the incident. You'll also need to understand what the hostage takers, negotiators, and possible rescuers are thinking and feeling.

By understanding the various psychological attitudes and motivetions of the involved parties, you will be able to control your own emotions, to think objectively with a clear mind, and to use the interplay of all these emotions to your own advantage.

When the incident begins, regardless of the type of hostage situation, the first emotion you are likely to feel is utter disbelief. *This couldn't be happening to me!* you'll think. This disbelief is a psychological coping mechanism known as denial—denying that the incident is real of which you are playing a part. You may be convinced that you're in no danger. This defense is common and effective but cannot be maintained for long. When denial does fade, the hostage has to face the danger to his life.

As soon as denial starts to fade, mind and body-numbing fear begins. Often a hostage will feel so powerless, shocked, and frightened that a temporary paralysis sets in. Escape attempts at this point will not likely be successful. The time for an early escape may only last the first ten seconds of the incident, before the hostage takers are completely in control. Once the incident is firmly underway, it is better to wait for a safer psychological and physical environment.

When this stage is reached, expect to feel helplessness and hopelessness, and disorientation. Often thoughts of escaping happen during this period, but the hostage can't make his body or mind cooperate. Don't attempt an escape during this phase! The likelihood of a fatal hesitation is too great. WAIT!

One of the stranger common occurrences in hostage situations is a regression by the hostage to his childhood. Being in a hostile environment, being isolated, and being helpless causes the hostage to forget past adult experiences and to resort to early adaptive behavior patterns from childhood. One female hostage found herself weeping constantly during an incident because she always got what she wanted from "Daddy" by crying nonstop.

The Developing Hostage Situation

There are several types of hostage situations. Kidnapping is a tactic used primarily by terrorists and criminals as a means of obtaining money through ransom. Kidnappings have other benefits too, such as:

- providing publicity
- obtaining the release of jailed comrades
- forcing large foreign corporations which have been frequently hit by kidnappings to leave the country, and
- forcing a government into granting concessions.

The major difference between a kidnapping and any other kind of hostage taking situation is that the kidnapper hides his victim while the other hostage takers confront the authorities with their victims.

Kidnapping requires extensive planning and precise execution in both the seizure and holding phases, which we will discuss in more detail later. The two primary methods that kidnappers use to accomplish a kidnapping are to take the victim from a static location, such as the victim's office or residence, or to kidnap the victim enroute, either when he's on foot or in a vehicle.

Static Location

In order to take the victim from a static location, the kidnapper must know: the location; what security measures are employed; what time the victim is at the location; how aware the victim and his family, friends, and associates are; and the number, times, and routines of police and/or guard patrols in the area.

To gain this information, the kidnappers will survey the location and area and make dry runs to find the best way to penetrate the location. The kidnappers will use ploys and ruses to penetrate the location. Ploys and ruses used in the past have included:

- repairman
- salesman
- survey-taker
- policeman
- mailman
- deliveryman
- business appointments cleared for access
- requests to use the phone
- requests for information
- and surreptitious entry

A single individual posing as a meter maid reconnoitered Brigadier General Dozier's home. If Mrs. Dozier had only been aware of the utility company's policy of sending meter readers out in pairs, she may have become suspicious.

The actions for a kidnapping from a static location include:

- penetrate the location
- gain control of the victim
- remove the victim from the location
- and escape without being detected

Brigadier General Dozier's kidnapping was a classic static site operation. Terrorists entered his quarters posing as plumbers. The story they used to gain entry was they were checking a leak in the apartment below his and wanted to see if the leak was originating from his apartment. General Dozier believed their story and let them in. The terrorists gained control of General Dozier by force and by threats against his wife. He was drugged, placed in a large box, and carried out of his quarters. The terrorists escaped with General Dozier in a van, leaving Mrs. Dozier bound and gagged, locked in the laundry room to ensure they would escape undetected.

Enroute
In order to kidnap a victim enroute, the kidnappers must know the victim's routine. Once a routine has been established, the kidnappers can plan an attack to include a time and an attack site. The steps in kidnapping a victim enroute are:

- select and identify the victim
- gather intelligence concerning the victim's vehicle and/or travel routines (to include times and routes)
- select an attack site along the route
- and select a time of attack based upon the victim's routines and the patterns of the police's security patrols

It's fairly simple to take a victim on foot enroute, once his routines are determined so that a day, time, and place can be selected. The kidnappers will then gain control of the victim, throw him into a vehicle, and drive away to the holding area.

The scenario for taking a victim from a vehicle is:

- isolate the attack site and stop the vehicle
- gain control over the victim
- and escape

Every kidnapping incident in which the victim is taken from a vehicle follows this scenario. Any difference will come in the ploy or ruse followed by the kidnappers to isolate the attack site and to stop the vehicle. Ploys and ruses used previously to stop vehicles include:

- roadblocks
- accidents
- hitchhikers
- ramming
- police disguises
- and by pushing bicycles or baby carriages in front of the vehicle

To isolate or secure the attack site, kidnappers have called security forces prior to the attack to tie up the phone lines, used fake police roadblocks and accidents as diversion tactics. A classic example of a kidnapping while enroute in a vehicle was the case of German industrialist, Hans Martin Schleyer. In kidnapping Herr Schleyer, the Red Army Faction used roadblocks and telephone calls to isolate and secure the area. A stationary vehicle was used to channel Schleyer's vehicle and the security chase car. A woman rolling a baby carriage and a moving taxicab were used to cause Schleyer's chase vehicle to pin his car in.

Pre-Incident Preparations

The criminal hostage taker (excepting the kidnapper) generally does not have enough time to prepare prior to the hostage incident. Most of these scenarios evolve as happenstance situations with little or no previous planning. Terrorists and kidnappers, however, usually go through elaborate preparation rituals prior to their taking of hostages.

In this pre-incident phase, the would-be abductors determine what they hope to gain from a hostage situation—ransom, political concessions, media attention, or whatever. Terrorist groups are especially professional in their approach to incident preparation. Preoperational activities are meticulously planned. Reconnaissance missions are conducted against the intended target (the potential hostage) and/or the areas of future operations by small, special intelligence teams. The leaders of the terrorist group then conceive and prepare the detailed plans for the upcoming operation based upon the information gathered by these special teams.

For security reasons, often planners, reconnaissance teams, and the actual abduction team will never meet. Information and orders are passed along through intermediaries, liaison sections, or by message drops. Training and rehearsals sometimes take place in countries outside the target area. Perpetrators, even the leaders, often have no knowledge of which specific target will be taken until it is time to conduct the operation. If a primary target is unavailable, or the risk is perceived as too great, an alternative target is selected. Most terrorist contingency plans include alternate targets. The plan may also include alternative negotiation demands and departure or escape routes.

Although most criminal kidnappers do not have this much organizational support, the real pros will go through a similar process within the scope of their particular capabilities. The kidnapper, however, will not generally have alternative victims.

Phases of Hostage Situations
There are four phases of hostage incidents acknowledged by most subject area experts:

- the capture
- transport/consolidation
- holding
- termination

Capture Phase
In terrorist situations, the strike team is brought together and briefed on the specific primary and alternate targets. They conduct final rehearsals to fine-tune the operation, move to the attack site, and then carry out the plan. Once commenced, the action is at the point of no return and will continue on through to its natural consequence. Kidnappings by criminals start similarly but with less emphasis on security and support.

Ordinary criminals and emotionally disturbed hostage takers generally spend little or no time on preparation but find themselves taking hostages as a matter of necessity or inspiration rather than as the natural outcome of some elaborate plan.

Transportation/Consolidation Phase
This is also situation dependent. In some cases, the hostage takers have enough freedom to transport their victim(s) to a site they deem adequate as a confinement area. In other cases, the hostage takers are forced or choose to make do with what they have and remain at or near the capture site, setting up a defensive perimeter. In other words, a terrorist group may ambush a dignitary, throw him into a car, and transport him to a hideout. Or, a criminal may take over a liquor store and remain there with whatever customers and clerks he was able to take hostage—

holding out against the authorities outside the premises. He consolidates his position so that he is less vulnerable to the police surrounding the outside of the building.

Holding Phase
After the hostage is either moved to a holding location or after the captors have consolidated their position, the incident enters what is its most lengthy phase—holding. In many ways, this is the safest phase for the hostage. The situation has had a chance to stabilize. There is little or no direct action taken by the authorities. Finally, a psychological phenomenon called the "Stockholm syndrome" has an opportunity to develop. The Holding Phase is a waiting period and is filled with negotiations.

Termination Phase
Regardless of how a hostage-taking incident ends, the Termination Phase is generally a very tense time for all people involved in the crisis. The easiest and safest termination for the hostage is a voluntary release followed by a hostage taker's surrender. More dangerous is the escape, and most dangerous is the rescue by an outside force or the killing of the hostages as a means to end the crisis.

The Stockholm Syndrome
The regression to one's childhood mentioned earlier sets the stage for another common phenomenon known as the "Stockholm syndrome", so named after a bank robbery hostage situation in Sweden during the summer of 1973. Jan-Erik Olsson, a professional safecracker and thief, recently escaped from prison, walked into the Sveriges Kreditbank of Stockholm brandishing a submachine gun. In the next six days, Olsson managed by means of the three women and one male hostage he was keeping, to hold the police at bay and to obtain the release of a former prison mate, Clark Olofsson. He eventually walked out in surrender with his hostages voluntarily clustered around him to protect him from police snipers.

The hostages had come to identify with their captor in a positive manner. One of the women even wanted to marry him. On the other hand, they were angry at the authorities for prolonging their captivity and not giving in to all of Olsson's demands for safe conduct to freedom.

The syndrome has three components:

- Positive feelings on the part of the hostage toward the hostage taker.

- Negative feelings on the part of the hostage toward the authorities and rescuers.
- Positive feelings on the part of the hostage takers toward the hostages.

The regression to childhood feelings and actions by the hostage to the hostage taker sets up a striking similarity between the relationship of the hostage taker and that between an abused child and an abusive parent. The child clings to the abusive parent just as the hostage clings to the hostage taker. The abused child and the hostage with the Stockholm syndrome are strikingly similar in that both are loyal to the "parent" out of fear. Both feel threatened by intervening authorities and have a tendency to defend a cruel "parent". The gun of the hostage taker becomes the instrument that demands loyalty.

Hostage fear and anger toward authorities and rescuers are the result of concerns that their actions may cause the hostage taker to become violent, taking out his anger on the hostages. These negative feelings are exacerbated by the concern that the negotiator is stalling, thereby prolonging the unpleasantness of the incident. It is not uncommon for hostages to become more afraid of the police than they are of the abductors.

If you find yourself straining to "understand" your captor's viewpoint and striving to agree with it, you are experiencing Stockholm syndrome. Be aware of its development, because you can expect the hostage taker to possibly reciprocate. There's a feeling of "we're all in this together".

It is not unusual to see the beginnings of this phenomenon after the first hour or so. Showing pictures of your family may help coax the abductor to see you as a fellow human as opposed to a faceless victim. You'll want to keep the incident as human as possible. It is when a hostage taker puts a bag over a hostage's head and face to make him seem less human, that the possibility death is increased. It's harder for a hostage taker to kill or harm someone with whom they have started to identify.

The hostage incidents in which the Stockholm syndrome has not been a factor have usually involved constant abuse from the hostage takers. Where constant threat of life and torture are involved, there will seldom be a development of Stockholm syndrome.

In your anxiety, you may find yourself angered by any actions made by your fellow hostages that may disturb your captor. This is valid from the perspective that you don't want the boat rocked if it is going to endanger your safety. Also, it's easier and safer to transfer your anger at being held onto a fellow hostage than it is to express anger toward your captor.

Hallucinations

You may find yourself experiencing unusual feelings and visions brought on by the stress of a hostage situation. If you have been isolated, confined, restrained of movement, and subjected to life-threatening danger, you may experience hallucinations and claustrophobia. We know the phenomenon doesn't depend on who takes you captive. The consistent factor is the combination of conditions mentioned above. If you are kept tied up in a small, dark room or closet by yourself and are threatened with death by your captor, you may have one or more of the following unusual experiences:

- Sensitivity of your eyes to light and difficulty in focusing your eyes.
- Flashes of light and geometric forms.
- Sensations of being in a tunnel, hallway, elevator, dark alley, corridor, or funnel.
- Detachment from what's going on—out-of-body experiences.
- Childhood and other past experiences relived in great detail.
- Voices and other auditory distortions.

Your mind is accustomed to constant stimulation. When your mind is deprived of its stimulation, it will manufacture its own. This does not mean you are going crazy, it only means your mind is coping with a bad situation. Once you are released, you may have flashbacks. Understand them for what they are—harmless perceptual accommodations.

The Hostage Takers' Point of View

At the start of a hostage incident, the hostage taker is at his most dangerous state of mind. He too is frightened—afraid of the consequences of failure, afraid of being hurt or dying, afraid of losing his freedom, and afraid of losing control. His nerves are on a hair trigger. You may scratch an itch and find yourself getting shot for your movement, which he interpreted as threatening. Keep your eyes averted. Do not challenge him or argue or question his commands in any way at this point.

Later on, when the Stockholm syndrome has had a chance to develop, you may try some of these things. By that time, he will have calmed down a little and should be more reasonable. The most important things for him at the beginning of the incident are to assure his own safety and to gain complete control of the situation as quickly as possible. Those who try to thwart him in any way or just appear to do so put themselves and others at great risk. Hang tight for the moment!

You should understand that not all hostage takers are alike. Some are less dangerous than others. The hostage taker who is a genuine political terrorist is unlikely to be capricious or irrational. He is also unlikely to be affected by appeals to personal selfish interests: that is, he cannot usually be bought off. He is conscious of the high risks he takes in his exploits, but does not protect himself from danger in the way that either a criminal or a policeman would. Yet for all that, he is a professional. He is dedicated to his job and, though he may want very much to avoid dying, he knows that he may have to.

The professional criminal, on the other hand (with the exception of kidnappers), will normally only take hostages as a last ditch effort to stay free. Given time to think about his alternatives, he'll give himself up rather than harm the hostages and earn an even longer prison term. A kidnapper, however, is very dangerous. Once he has gotten you to confirm that he has you and that you are still well, your value decreases. Many kidnappers kill their victims to prevent later identification. Terrorists, on the other hand, may kill you out of revenge.

Flight
Escaping from captivity is a personal decision. You must weigh the possibility of escaping against your escape failing thus resulting in tragedy. Just as important is your evaluation of your chances for survival and ultimate release. Mental/emotional and physical capabilities to execute the escape must be assessed. If you determine that you are capable of executing an escape and that your chances for survival and freedom are good, you must plan logically and in detail for all contingencies. Given proper planning, a thorough assessment of both yourself and your captors, and a strong desire to survive an escape to freedom, you will have the best possible chance for freedom should an opportunity for escape, release, or rescue arise.

A hostage planning a getaway must have a plan for continuing his escape, which will take him out of the area of captivity. This takes him out of the abductors' immediate search area and provides him with an objective to strive for that maintains his present freedom and eventually provides his ultimate freedom and safety.

Planning for the Abductor Release

The most important thing you can do in anticipation of this eventuality is to constantly observe your captors and their activities and habit patterns. Memorize everything you can about them, so that you can pass on as much usable information to the negotiating authorities as possible. This also includes information about the containment area, which could prove critical to an assault team's plans. Remember, you may be the only source of inside, up-to-date knowledge about what the authorities are facing. Try to help them help others.

Planning For an External Force Rescue

The hostage has no direct control over a rescue by an external force. Additionally, in most cases of external force rescues, you will have no prior notice of the impending rescue attempt. Quite often hostages become the innocent victims of their rescuers when the rescue force confronts the captors and a shoot-out occurs. Martin and Gracia Burnham are excellent examples of that. You can, however, plan certain passive and active measures, which may save your life during a rescue confrontation.

Passive Measures

Passive planning should first include the assumption that your abductors won't release you and you cannot escape. A second assumption should be that once your location is known, a rescue attempt would be made. Based on these initial assumptions, certain passive measures should be planned and practiced. The first of these is to make it a point to always sit or lie down when in your captors' presence, unless they don't allow you to do so. You should also try to keep as far away from them as possible.

These actions alone could save your life during a rescue attempt by keeping you out of the line of fire between the captors and the rescue force.

Another passive measure that you should attempt is to maintain a cordial and respectful relationship with your captors. The establishment of a human bond between hostage and captor may deter the captor from executing the hostage when he realizes a rescue is taking place. It is much harder to murder someone with whom a friendly bond has been developed than it is to murder someone who argues and is defiant. Several Jewish hostages during the Entebbe Raid owe their lives to a German terrorist with whom they had earlier made to feel ashamed for his countrymen's actions during the holocaust. When the Israeli Commandos stormed the airport buildings, he looked at them as if ready to execute them. One of the hostages asked him if he was like the

Nazis. That reminder was enough for him to turn from them and defiantly go down in a blazing gun battle with the Commandos rather than to harm his charges.

A final passive measure is to always be alert for signals that a rescue attempt is about to take place.

Active Measures

Planning active measures to take during a rescue attempt is extremely important. The first action a hostage should take upon realizing that a rescue attempt has commenced is to immediately lie face down and attempt to remain motionless and silent. You should not try to physically or verbally make contact with the rescue force. Any such attempt could result in your being mistaken for a resisting captor and getting shot. You may also draw vindictive attention from your captors. You should allow yourself to be apprehended by the rescue force and taken from the place of captivity. There will be time later to identify your self as a former hostage and have that identity verified.

The passive and active measures identified above are simple to plan and easy to implement; however, hostages have been injured or killed more frequently than necessary simply because they failed to plan for and implement these measures for their own safety. Such a small effort for such a big price to pay!

Flight

There may come a time during your captivity when you might consider using violence to make your escape. If you are not well trained, if the timing is wrong, if you are not committed to following through on your actions, you may get injured or killed, or you may cause others to be hurt or killed.

You must know your limitations! If you are older and/or not in good shape, you need to understand that your reflexes will not be as fast or as smooth as you remembered them when you took that karate course 20 years ago. Pride lets us see ourselves in the best possible light. That could be lethal!

The hostage takers may be younger than you; he or she may have had a recent intensive course in hand-to-hand combat; and they are more emotionally hyped thereby increasing their speed with adrenaline. If they have initiated the taking of hostages, they may have planned out every aspect of it and are expecting trouble while you have no forewarning and can only react. They may be more desperate, considering they have nothing to lose.

You must consider these and many other aspects before you attempt physical violence! If there is the slightest chance of hesitation or fumbling on your part, you may be better off not trying anything! If the captors are Middle Eastern, however, go for it. They're going to kill you anyway.

Consideration of Fellow Hostages

If you deem it necessary to use physical violence (for instance, the captors have already killed or wounded others for seemingly no reason and they seem about ready to do it again), first consider the possible impacts your actions may have on any other fellow hostages. Will they be inadvertently harmed by your actions or through the captors' retaliation after your attempt, whether it is successful or not?

If there is a gun involved, might it discharge in the direction of innocent people? If there is a grenade or bomb involved, is it worth wrestling with the captor? Is there a safe place to throw it should you gain control of it? How long a delay will there be once the firing mechanism has been triggered?

There's more involved than one might think when weighing the chances of successfully overpowering a captor or a guard. For instance, when this author trains security guards and Military Police in foiling assassination attempts, he uses disarming techniques that turn weapons such as guns back into the wielder, so that only the attacker is endangered in the fray. If you feel physical resistance techniques are called for, please consider the safety of innocent bystanders.

When and If To Use Violence

Escaping a hostage situation is similar to escape in a prisoner of war situation. One must do it immediately during the capture process, while a high state of confusion exists. Or, you should wait until later after there is time for careful planning and when the guards are possibly less vigilant. It is common for terrorists to kill or injure someone soon after the incident begins to prove they mean business. If a chance to use physical resistance comes early on, you might first want to reconsider your capabilities in comparison with those manifested by the hostage takers. If you don't, instead of saving the guy they were going to make an example of, they'll make *you* the example. In the case of a criminal, you might want to hold back because they don't generally kill right away since they need hostages for self-protection. The terrorist, on the other hand, is more interested in revenge and will do just about anything to get it.

If the captor is systematically killing off people or if he appears insane and begins dehumanizing his victims by covering their faces with masks or hoods, you may be forced into physical resistance action simply out of self-preservation. As a general rule, however, a victim should remain passive and avoid eye contact with the assailants thus living to fight another day. No one admires dead heroes or bumblers who cause injury or death to themselves or other victims. Examine your motives before acting rashly.

You must also be careful in that there may be more opponents than you realize. Sometimes terrorists use a "sleeper" agent (a fellow terrorist) who watches the victims' initial reactions for signs of "trouble makers" before he shows his true colors. Be aware of this possible danger!

The safest time for physical resistance is when guards have been lulled into carelessness. On two different occasions, an American OSS agent was able to escape from his Gestapo captors by using a pencil to kill the lone Nazi interrogator left with him to obtain a confession. Since the prisoner had looked so helpless, no one assumed that he would attempt to overcome a lone, armed guard. How ironic that the guard actually handed the prisoner the very implement by which he lost his own life.

Working toward an End to the Situation
Try to help the authorities in their attempts at hostage negotiations and rescues. The best way your company can do that is by providing information on the facility where the hostages are being kept, in their personal information, and in your business's routines. If the hostage taker is an employee, any personal information on him is critical. Cooperate in way you can to make the authorities' jobs easier and more effective.

Your Worst Nightmare—
Dead and Wounded Employees and Customers
Express immediate concern and sympathy publicly and privately. Set up support mechanisms and groups for any survivors and their families. Provide professional counseling for them. Determine the economic impacts of the incident to all involved and try to come up with fair compensation for at least their immediate needs.

Coping with the Aftermath
Let's say you have been taken hostage and have been fortunate enough to be released, to escape, or to be rescued. How will you feel? What do you need to watch out for? What should you do?

Remember how a common reaction to being taken hostage is an initial sense of disbelief, as if this couldn't be happening to you. Now that it has happened, you may feel, "This really did happen to me. I'm now different than I was. Why doesn't everybody else realize that?"

We feel this reaction is similar to that experienced by many Vietnam veterans when they initially came back home. In their case, one day they were fighting in the jungle and heat, watching their buddies die and experiencing the mind-numbing terror of combat. The next day they found themselves strolling down their hometown main street.

They had a very deep sense of unreal feelings that none of their old acquaintances would ever truly understand them again. When a car backfired, they may have automatically dived to the ground with their combat survival reactions still in place.

As it was for the war veteran, so it can be with ex-hostages. A direct correlation can even be made with the pervasiveness and long lasting effects these experiences often have. Just as there are some Vietnam veterans who still find it difficult to deal with their past war experiences twenty years and more later, the ex-hostage may also find his terror and trauma lingering on. Even though the captivity experience may have only lasted a few hours, the aftereffects may hang on for years. Let's take a look at some of these post-incident reactions and what can be done to return to a life of normalcy.

The Stockholm Syndrome Revisited

As mentioned before, it is common to become emotionally attached to the hostage taker during the incident. It is also common to fear or become angry at the negotiating authorities and rescuers because they are perceived as life threatening to the hostage. After the incident is over, it is not uncommon to feel guilt, shame, and puzzlement over these feelings. Indeed, these post-incident reactions are so common that they have become known as the fourth phase of the Stockholm syndrome.

Common Problems

There are a number of problems that frequently occur after the hostage incident is over. Some of these are physical in nature:

- Loss of appetite
- Unexplainable episodes of dizziness
- Sexual dysfunctions
- Sleep disturbances and insomnia

Many of the problems; however, are psychological:

- Being startled easily
- Shame and guilt
- Decreased motivation
- Claustrophobia and other phobias
- Anxiety caused by decision-making
- An erratic or changing temperament
- Nightmares and recurring dreams of the incident
- A feeling of being cheated out of life experiences
- A feeling of numbness or detachment
- A sense of entitlement
- Withdrawing from former interests and relationships
- Difficulties in memory or concentration
- Avoidance of "Triggers"
- Unwillingness to be "Treated" by a counselor

What Helps

After seeing all the physical and mental problems brought about by having been a hostage you might wonder what's the use? Why go on living in misery? We think, however, that you are the one element in this equation that can really make the difference. You might have been helpless during the hostage situation, but once you're out of it, you have the right, the capability, and the responsibility to control your life and what goes on in it! What you do counts so make it meaningful! These are fine sounding words, but how does one go about recovering from the aftereffects of being a hostage?

First, you need to expect problems and be willing to meet them head on. No one ever said it would be easy. You must be willing to try to over-come any problem that rears its head. Don't allow yourself the luxury of remaining helpless after the incident. That would be a cop out!

Second, you'll need to talk out your feelings. See a properly trained therapist and heed the advice and help you are given!

What the Company Can Do

Help the victim uncover the specific events in the hostage situation, which he has been attempting to deny or which he has been unable to accept.

Support him as he feels the feelings associated with these traumatic events.

Help him acquire a greater understanding of how his hostage experiences have affected his life in the present.

Help the victim find constructive uses for his hostage experience.

What the Victim Can Do to Help Himself
Get back into a routine as quickly as possible. Routines are natural sedatives in a manner of speaking. Let's face it, if you've just come back from being a hostage you've probably enjoyed about as much excitement as you can stand. Settle back down into the humdrum of day-to-day life. Use this as an anchor to reality.

Finally and most importantly, emphasize the positive. Try not to dwell on all the bad things that happened or may still be happening as aftereffect. Think rather of those things about yourself and others that have grown and matured from this experience.

The Need for Planning and Standing Operating Procedures
The time to prepare for hostage situations is now—before one happens. Careful planning and developing routine procedures based on those plans are imperative to give your company or organization the best possible chance of handling crisis situations like these.

The Need for Employee Training
Employees and management need to become familiar with portions of those plans and practice how to respond if a hostage situation ever develops in your company.

The Need for Insurance
Some of the new Homeland Security laws and regulations have made this a controversial area lately. Check with your insurance agent about the availability of terrorist and hostage riders on your company's liability policies. Protect your company's interests with life and ransom policies on key personnel.

Need for Public Affairs Management
A hostage situation will focus the eyes of the media upon the situation and how your company deals with it. Make sure you think through how such situations will be handled with the media. Prepare for damage control now, not when you're surprised by an incident in the future.

Conclusion
Hostage situations are a grim fact of life today. The future promises

they will become even more commonplace. Think through the possibility and prepare now!

Much of this material was taken from Mr. Spear's book "STAY ALIVE: Survival Tactics for Hostages." Mr. Spear is a retired military intelligence professional, a business owner, and a 7th Dan in Hapkido. He has trained over 11,000 people in self-defense and personal security since 1974 throughout the world. He has had nine books and five videotapes published in these fields.

Chapter Four

Fraud and Theft

Terrorist Links to Commercial Fraud
By Peter Goldmann

For years terrorists have used commercial fraud in the United States to raise funds for their deadly activities.

While the September 11 terrorist attacks are far from being fully investigated, there are already shreds of evidence pointing to the involvement of various forms of fraud in the World Trade Center and Pentagon attacks.

Here are some of the schemes known or suspected to have been connected to the September 11 attacks. They have red flags that businesses, governments, and law enforcement agencies can screen for to find would-be perpetrators of future terrorist atrocities.

Terrorists' Favorite Frauds

- Grocery coupon fraud. This is one of the terrorists' favorite fraud schemes for financing their operations.

    Example: The 1993 World Trade Center bombing. This act was financed in large part by an elaborate coupon fraud scheme. Mahmud Abouhalima, Ibrahim Abu-Musa, and Radwan Ayoub established a network of grocery stores in New York, New Jersey, and Pennsylvania. Abouhalima was one of the four suspects sentenced to life in prison for his involvement in the 1993 bombing.

    How it works: Typically, the terrorists set up "cutting houses" where women, children, and college students spend hours clipping coupons stolen from manufacturers' plants, print shops, or newsstands. They are purposely wrinkled so they appear to have moved through legitimate retail grocery channels and are shipped in bulk to industry clearinghouses, which in turn issue redemption checks.

    Red flag: A small corner grocery, which normally redeems $200 to $300 a month in coupons, suddenly starts turning in thousands of dollars worth. Coupon clearinghouses should not hesitate to check these cases out.

- Export diversion and money laundering. Here is an example of how it works: ABC Corp., a major American manufacturer of electrical appliances, sells widgets in the United States and Europe for $10 each.

In Third World countries, however, ABC will sell the same item for just $5 apiece by "stripping" away the normal distribution, advertising, and promotional costs.

On March 1, ABC receives a $500,000 order from XYZ Trading Company Ltd. of Mozambique. XYZ states in its letter and in its subsequent telephone calls with ABC that the order is for product to be distributed in Mozambique.

ABC replies with an offer to sell 75,000 widgets to XYZ at the price offered—$6.50 each—with the specific understanding that the goods are to be sold in Mozambique and nowhere else.

ABC specifies that the terms are either cash or irrevocable letter of credit (L/C). Again, XYZ agrees. A confirming L/C is opened through a Swiss bank to the credit of ABC's account in an American bank. ABC delivers the widgets to the Port of New Orleans. It presents the appropriate shipping documents to the correspondent bank in the United States and the L/C is paid. As far as ABC is concerned, the deal is completed.

Minor detail: Because there are not many direct shipping routes to Mozambique, XYZ asks ABC to ship the goods to Rotterdam where XYZ will take care of the final leg. Upon arriving in Rotterdam, the containers carrying the goods are off-loaded from the ship and transported to a warehouse where they are stripped and the products re-stuffed in new containers.

The new containers, rather than being shipped to Mozambique, are put on the next available ship to the United States. Upon arrival in the U.S. port, the goods are declared as "U.S. Goods Returned" through U.S. Customs, thereby avoiding any U.S. duty.

After passing through customs, the goods are stripped from their containers and immediately shipped to U.S. stores for re-sale. The U.S. importer—QRS Corp.—has a direct, though carefully disguised, relationship with XYZ.

Result: ABC has no idea of the identity of the ultimate consignee (QRS). QRS sells the widgets to a "fence"—a whole-saler who may be perfectly legitimate—for $8.25 apiece. Everyone is happy. The wholesaler gets the widgets for $1.75 less than the regular market price, and QRS makes a compareable profit over the original purchase price.

Key: The money for the initial transaction often originates in countries such as Panama, Hong Kong, Colombia, and Pakistan and is then routed through banks in the Caribbean and Switzerland.

If the U.S. "importer" (QRS) is asked about the source of its funds by government investigators, it can respond that it earned the money by selling imported merchandise and has the documents to prove it.

The scheme thus has both earned a profit for the "importer" and has served as a channel for money laundering.

The Terrorism Link

Investigators tell the magazine *White-Collar Crime Fighter* that they have learned of numerous direct connections between fraudulent transactions such as these and terrorist groups in the U.S. and abroad.

Private sector role: For U.S. manufacturers, it is essential to apply rigorous due diligence procedures on any sale to a new customer—especially those in Third World nations. In the example above, the red flag for ABC should have been the request of a first-time customer to ship the goods to Rotterdam, rather than Mozambique.

Essential to effective "Know Your Customer," research is used from the many government and private-sector databases listing individuals and organizations that have been identified as "denied parties," with whom it is illegal to transact business.

According to Larry Christensen, a Vice President of the Dulles, VA-based international trade services firm, Vastera, there are 140,000 names on U.S. government blacklists, of which about 850 are alleged terrorists.

To be sure you're not negotiating an export deal with a terrorist-linked organization, check one or more of these databases:

- Bureau of Export Administration's "Denied Persons List," which includes names of individuals and organizations that for various reasons have been denied export privileges from the U.S. www.bxa.doc.gov/DPL.

- U.S. Treasury Office of Foreign Assets Control, which can be accessed at www.treas.gov/ofac. This database contains an incredible amount of detail about hundreds of terrorist organizations, as well as individual terrorists with whom it is illegal to trade. It is the list to which the White House on October 12 added 39 names of people and organizations whose assets were ordered frozen in connection with the World Trade Center and Pentagon attacks.

- Trade Compass at www.tradecompass.com, which combines several key publicly available lists of blacklisted entities.

- The Washington, DC-based law firm deKieffer & Horgan maintains a comprehensive database of known export diverters, counterfeiters, smugglers, and their accomplices, in addition to files on companies and individuals identified as terrorists (or their surrogates) by the U.S. government. Ddekieffer@dhlaw.com

Track your products. Detecting an illegal diversion of your goods at an early stage is essential to avoiding legal problems. There are several reputable companies that can help with packaging and marking technologies for tracking your shipments. A reputable export consultant can provide leads.

If you do discover a cargo diversion, or you suspect that you have inadvertently done business with a suspicious company or individual, immediately contact your attorney as well as the FBI and the U.S. Customs Service.

Other Frauds To Watch For
- Credit card/check fraud. The Secret Service has reported instigation by groups of Middle Easterners with known terrorist ties of so-called "booster check/bust out schemes."

    Recent example: The perpetrators, organized into "cells" located throughout the U.S., applied for and received numerous credit cards. In one case, some members received in excess of 40 credit cards.

    The cardholders systematically "boosted" the credit limits to twice or even three times the original credit limit. This was accomplished by sending worthless checks to the issuing bank in excess of the original credit limit, thereby creating a credit balance and effectively boosting the account's limit.

    The check amounts were posted to the accounts prior to the checks clearing the bank.

    Before the checks were returned as worthless, the cardholders purchased merchandise and/or obtained cash advances, up to, and in most cases, in excess of the limit on the accounts, thereby "busting out" the accounts. Losses to banks and merchants from these frauds exceeded $4.5 million.

- Solicitations by phony charities. This activity has been widely reported. Legitimate-sounding names like Wafa Humanitarian Organization and Al Rashid Trust are among the many notorious fronts for Muslim terrorist fund-raising.

  This is related to so-called "reverse money-laundering" where legitimate businesses (controlled by terrorist sympathizers) give generous donations to terrorist organizations. The money goes from legitimate sources to "dirty" donees, rather than the other way around.

- Identity theft. Some of the hijackers in the September 11 attacks were known to have fraudulently used the identities of others, including in one case that of a Saudi individual who had been dead for several years.

- Document forgery. The September 11 hijackers carried false driver's licenses and other phony ID. While it's hard to screen for these documents, poorly produced forgeries do often stand out. In these cases thorough verification procedures should immediately be applied.

White-Collar Crime Fighter sources: Todd Sheffer, CFE, Trade Intelligence LC, Mt. Airy, MD, tilc@adelphia.net.
Donald E. deKieffer, deKieffer & Horgan, attorneys, Washington, DC, www.dhlaw.com.
Peter Lilley, Proximal Consulting, Marlborough UK, www.proximalconsulting.com.
Federation of American Scientists, Intelligence Resource Program, www.fas.org/irp

Peter Goldman is the Publisher of Cyber-Crime Fighter, a monthly newsletter for corporate and law enforcement computer and Internet security professionals. He is also the Course Developer of FraudAware, a first of its kind Web-based employee fraud-awareness training course.

Insurance Fraud

By Barry Zalma, Esq.

Working to defeat fraudulent insurance claims is often a thankless task. After a thorough and comprehensive investigation the insurance fraud professional will often be faced with the following disturbing developments: instructions to pay the claim and close the file because "you have too many pending claims"; and instructions to return the claim file back to the claims adjuster. He is then faced with:

- Insured or claimant who files suit so the file is turned over to defense counsel to defend.
- Defense counsel with no experience with fraudulent claims who defends the suit as if it is a simple bodily injury claim.
- Prosecutors who believe an insurer cannot be a "victim" of any crime.
- Judges who believe an insurer can never be a "victim" but is always the aggressor.
- Suit filed against the company and defense counsel assigned.
- Defense counsel who ignores the insurance fraud professsional.
- Court biased against insurers who refuse to recognize legitimate defenses.
- Insurer compelled because of the actions of the court, or its own counsel, to settle with the plaintiff rather than bleed to death paying attorneys fees.
- Jury that ignores the facts and rewards the fraud perpetrator.

All of these events happen, or will happen, on thoroughly investigated fraudulent insurance claims. Insurers lose to fraud perpetrators even when the facts and the law are on their side. Insurers lose when they are represented by the best, most knowledgeable and experienced attorneys. It is frustrating to lose to a fraud perpetrator. Losing can cause insurance fraud professionals indigestion, gastritis, and bleeding ulcers. More than anything else, it destroys confidence in the ability to gather evidence to prove fraud.

Insurance fraud professionals understand that they often will not succeed. They understand that no matter how thorough the investigation is, regardless of the superiority of the legal analysis of counsel for the insurer, and no matter how well the case is presented to a judge, the insurer will lose. Courts of Appeal seldom reverse the factual findings of a lower Court. When an insurer loses a bad faith case, its greatest hope is that the Court of Appeal will reduce the award.

Why do insurance fraud professionals keep on working against insurance fraud when they know they will lose some of the really good cases? Insurance fraud professionals continue because they know what they do is important. They know that a viable and effective insurance industry is essential to a vibrant economy. What insurance fraud professionals do keeps the wolves from the doors of the insurance industry. Insurance fraud professionals do what they do because they recognize that insurance is essential to the operation of American business. If fraud is allowed to succeed on a regular basis, the insurance industry and the economy can be destroyed. They do so in the face of ridicule from their friends, neighbors, and colleagues.

An insurance fraud professional who effectively deprives a person of the insurance benefits to which the insured was not entitled is not liked. The fraud is unhappy because he did not get the money he hoped to steal. The insurance broker is not happy because he has been branded as the person who sold a policy to a fraud. The company is not happy because the investigation cost a lot of money, and the file is still open. The insurance fraud professional can only be happy because of a job well done.

I am not writing to complain. Rather, I wish to praise those of you who are part of, or keep on salary, a staff of insurance fraud professionals. I also write to suggest a modification in current law.

Praise For The Insurance Fraud Professional

The insurance fraud professional is an investigator, a claims person, a claims supervisor, a Special Investigative Unit (SUI) investigator or manager, a forensic accountant, and/or a lawyer who spends a majority of her or his working life working to defeat perpetrators of fraudulent insurance claims. They have experience, skill, training, and dedication that are reflected in multiple hours beyond the basic eight-hour day. They work within a legal system designed to thwart their efforts, and most of the time they overcome that handicap.

I give my best wishes to each insurance fraud professional. I honor your work as your employers and clients should. Your work may not always seem to provide a financial benefit to the insurer for whom you work. Often, the costs of investigation and defense far exceed the amount it would have cost to pay the claim as the fraud perpetrator demanded. The insurance fraud professional and the professional insurer for whom he or she works resist the temptation to pay a fraud. The insurance fraud professional understands that fraud must be resisted with vigor regardless of the expense. They do so because they know the insurer will save money in the long term by deterring others from attempting fraud.

I can quantify from my personal experience that in the long term an aggressive and effective anti-fraud program saves insurers money and that when a fraud perpetrator is paid after a vigorous effort to defeat the fraud, even if he or she is paid punitive damages, the long-term results are to the benefit of the insurer.

Those who commit frauds of opportunity—people with legitimate claims who just pad the claim enough to cover their deductible—are deterred from taking advantage of the opportunity because the insurer publicizes the success of its anti-fraud program.

The insurers with aggressive and effective anti-fraud programs will find that the total paid out for all claims—whether identified as potentially fraudulent or not—will reduce proportionately with the effectiveness of the anti-fraud program.

In my experience one client reduced its total cost per claim (expense and indemnity payments) by 30 per cent by requiring the insured, on every suspected fraudulent claim, to submit to an examination under oath. Often the cost of counsel to conduct the examination under oath exceeded the amount of the claim. The savings were not made on the individual claims, but on those claims that were never submitted because it became common knowledge among those who are professional fraud perpetrators that the insurers anti-fraud program made it too difficult to commit a profitable fraud.

An SIU trying to quantify the success of its efforts should not concentrate on savings on the individual claims it investigates. Rather, it should compare the total cost of all claims, both expense and indemnity, before the institution of the anti-fraud effort and the total cost of all claims after the institution of the anti-fraud effort. This method will allow the insurer and its SIU to quantify the effectiveness of its anti-fraud effort. It is not the savings, or cost, of an individual fraudulent claim that matters but the annual reduction in total claims payments for expense and indemnity. If the only change in the insurer's operation is the institution of an aggressive anti-fraud program, the SIU can take credit for the savings. The Insurance Fraud Professional recognizes that one of the most important aspects of an effective anti-fraud effort is to deter fraud perpetrators from attempting fraud.

A Proposal for a Change in The Law

Fraudulent claims are paid by insurers voluntarily because of fear, fear of a suit for breach of the covenant of good faith and fair dealing when a trier of fact may be tempted to award punitive damages against the insurer. Insurers are afraid that if they refuse to pay a fraud perpetrator he will sue for the tort of "bad faith" and may receive punitive damages. A $4,000 property claim may result in a $4 to $40 million punitive damages verdict. The fear of a punitive damages judgment because the insurer obeyed the law and attempted to protect itself from a suspected fraud perpetrator defeats the purpose of the insurer's anti-fraud efforts.

There is a need for a tort of bad faith. Insurers wrongfully and without cause breach the terms and conditions of the policy of insurance and willfully cause damage to their insurers. There are bad people running insurance companies just as there are bad people presenting fraudulent claims to insurers. The bad people need to be punished.

Everyone in the insurance industry recognizes that bad facts make bad law. Bad court judgments make awful settlements. If a court renders a major punitive damages award against an insurer, dozens of insurers who are defending fraudulent claims in the courts immediately rush to settle those cases for fear of being painted with the same brush.

It is the public policy of the state of California and many other states that it is important to fight insurance fraud. For example, the California Insurance Code provides:

1871. The Legislature finds and declares as follows:
(a) That the business of insurance involves many transactions that have potential for abuse and illegal activities.
(b) That insurance fraud is a particular problem for automobile policyholders; fraudulent activities account for 15 to 20 percent of all auto insurance payments. Automobile insurance fraud is the biggest and fastest growing segment of insurance fraud and contributes substantially to the high cost of automobile insurance, especially in urban areas.
(c) That prevention of automobile insurance fraud will significantly reduce the incidence of severity and automobile insurance claim payments and will therefore produce a commensurate reduction in automobile insurance premiums.

Similar statutes making up the California Insurance Frauds Prevention Act (Sections 1871 et seq.) make similar declarations against other types of insurance fraud from property claims to arson-for-profit. Regardless of the portentous declarations of the Legislature that compel all insurers in California and most of the states of the United States to institute an effective anti-fraud program, the availability of punitive damages for perceived "bad faith" conduct is an effective deterrent to the implementation of an effective anti-fraud program.

I propose, that if the legislatures of California and those of the various states that are serious in their concern about insurance fraud, they enact a statute that protects the insurance buying public and the insurers equally. The statute should read in substance:

DAMAGES FOR INSURANCE BAD FAITH. If a trier of fact in this state determines that an insurer has committed the tort of bad faith, it may only award damages in the following categories:

- Actual damages that are the benefits of the policy of insurance that would have been paid but for the bad faith.
- Interest on the benefits at the legal rate from the date of the loss entitling the Insured to benefits of the policy.
- An amount to reasonably compensate the insured for bodily injury or emotional distress proximately resulting from the tort of bad faith.

If the trier of fact determines that an insurer has committed the tort of bad faith intentionally or in conscious disregard of the rights of the Insured in order to deprive the Insured of the benefits of the policy of insurance, it may only award punishment damages equal to three times the damages awarded in the three categories above and reasonable attorneys fees, set by the trier of fact, as incurred by the Insured to obtain the benefits of the policy.

Such a statute would deter those insurers who act in bad faith and protect those who do not from frivolous suits. It would eliminate the fear generated by the potential of a punitive damages judgment. Insurers faced with a potential fraudulent claim can calculate the cost of fighting it and can then set aside the necessary reserves. Since punitive damage awards are by definition impossible to predict, treble damages will remove uncertainty from the fight against insurance fraud, will remove the deterrent effect of uncontrolled and uncontrollable punitive damages, will allow insurers to institute effective anti-fraud efforts, and will compensate the insured and the Insureds lawyer who is treated badly. Insurers who do not treat their Insureds with good faith and fair dealing will still be punished.

California, and many other states, have imposed Regulations to compel insurers to institute anti-fraud programs. The California Regulations, for example, state the purpose of an SIU as follows:

The purpose of a Special Investigative Unit is to assure the effective implementation of Sections 1871 et seq., of the Insurance Code (the California Insurance Frauds Prevention Act) or to detect and investigate on behalf of the insurer suspected fraudulent claims by Insureds or persons making claims for services or repairs against policies held by Insureds and to deter insurance fraud and thereby reduce insurance costs. The SIU shall organize the elements of the insurer's integrated, corporate antifraud strategy. The SIU shall cooperate with the insurer's claims handlers who are trained in fraud detection, as well as the insurer's legal personnel, technical support personnel, and database support personnel.

This section establishes the first requirement I can find where a business is told by the government that it must, as the victim of a crime, create an integrated corporate strategy against the crime the state knows is being committed against the business. Most businesses demand that the police protect them when they are the victims of a crime. Here, the state demands that the victim fight the criminal or be punished by the state and assess a special tax against the victims of the crime to set up a special police force and to pay prosecutors to prosecute the crime. No other category of crime victim is compelled to institute programs to fight the crime perpetrated against it and pay more than any other citizen, individual or corporate, for police protection.. Further, if the insurer errs in carrying out this mandate, or the trier of fact believes it has erred, it is compelled by statute to place its neck on the chopping block of "bad faith" suits and claims for punitive damages.

The Regulations also require that the SIU shall meet the following primary objectives through the use of the expertise of the SIU staff:

(a) the establishment of a systematic and effective method to detect and investigate suspected fraudulent claims and to provide for their appropriate disposition;
(b) the education and training of all claims handlers to identify possible insurance fraud through matching specific claims against patterns and trends indicating possible fraud and against specific red flags, red flag events, and other criteria indicating possible fraud;

Failure to comply with the Regulations allows the Department of Insurance to impose on the insurer a $50,000 fine or a loss of its certificate of authority. The Regulations, coupled with the court made tort of bad faith, are contradictory and self-defeating. The insurer is placed between the "rock" of the fine and the "hard place" of the potential punitive damages award.

Only by doing away with defining punitive damages for insurance bad faith as three times the amount that would have been recovered had the claim been paid may the purpose of the Fair Claims Practices Act, California Insurance Code section 790.03(h), and the Fair Claims Practices Acts of the various states and their SIU Regulations, be fulfilled. Insureds will have a chance to defeat fraudulent claims and consumers will be adequately protected from insurers who act in bad faith.

Mr. Zalma is an attorney who has emphasized for the last thirty years fraud detection and the defeat of fraudulent insurance claims. Martindale Hubbell has rated him "a.v.". He is a Certified Fraud Examiner and is on the faculty of the Association of Certified Fraud Examiners and Claim School. Mr. Zalma's law firm limits its practice primarily to the representation of insurers and those involved in the insurance industry. Areas covered include: expert testimony and consultation regarding claims handling, fraud & insurance coverage; examinations under oath; insurance fraud; insurance coverage analysis; and alternative dispute resolution.

He provides legal advice and assistance concerning insurance policy coverages and suspected insurance fraud. Because of the immediate response required by clients who need coverage advice, Mr. Zalma refuses all litigation assignments. As an expert witness Mr. Zalma testifies for both the plaintiff and the defense.

The full text of the FCRA as amended may be found on the Internet at: www.ftc.gov/os/statutes/fcrajump.htm. I strongly suggest you read the amended rules. They now affect surveillance on employees. If an employer has one of the new forms signed and on file, then and only then can you do a surveillance on that employee. If you do a surveillance and this form is not on file, then you, as well as the employer, can be sued.

Do Not Become a Victim of a Scam
By Kelly E. Riddle

As the owner of a private investigation company and having more than 20 years experience, I never cease to be amazed when people *allow* themselves to become a victim of a scam. Yes, even the most prudent person can be a victim of unscrupulous businesses, but many of us are victims of our own lack of research. Since the mid-1980s, records have been pouring into databases that are readily available for us to review. I have found doctors and attorneys that are not licensed, men that have been married to more than one person at a time, businesses with more than 150 lawsuits against them and banks that are not actually a bank. I am sure that the lady who married a man who was already married with several kids didn't think she could be a victim. The people that sent funds to an offshore bank had no idea that the bank was just a shell and a scam. I wish I had a $100 for every person that paid a contractor good money and never had the work completed.

The issue of privacy always seems to present itself anytime the topic of information research arises. As a reminder, we are taxpayers and our money is used to compile, store, and access these records. As taxpayers, we have the legal right to this information whether we want it to be open or not. More importantly, if we have a child in a daycare and suspect an employee of being a pedofile, we want to know the answer to that question. What about those of us who have a parent that you are considering placing into a nursing home? I am sure that we would like to know the way to research these facilities.

So if information is so readily available, why do scams still happen? Most of us have lives, jobs, and families that take up our valuable time. Spending time searching the millions of Internet web sites for the one that can actually help requires more time than most are willing to spend. Even if you find information on the Internet, who has time to go to the courthouse to review the file? The following websites will provide you with the tools you need to protect your assets:

Criminal Records

www.memphisdailynews.com provides online criminal records for 22 states. By searching this database, you will be able to find out if the person has been arrested and charged with any crimes in that particular state. The record will identify the date, case number, court, charge, and disposition of the case. This information will allow you to learn if the person has tendencies towards physical abuse, thefts, fraud or other criminal activity.

Lawsuits/Judgments/Liens
www.knowx.com provides on-line searches state-by-state or nationally. This site will allow the user to key in the person or company name and find out if they have a history of poor workmanship, being sued for bad debts, or failure to abide by contracts. If liens are found, they may be a result of failing to pay federal or state taxes, another sign that the company may be having financial difficulties.

Doctors and Health Facilities
www.healthgrades.com provides information and ratings on doctors and health care facilities. This site allows the user to search the credentials of a particular doctor and identify the types of drugs they can legally prescribe. The site will also provide information on hospitals and nursing homes, including the type of medicine they practice, the number of beds in the facility, the emergency care they have available, and their overall ratings.

State Specific Searches
www.pac-info.com provides records searches at the state or local level by clicking on the particular state. The records available for each state differ, but most have access to property records, state corporate records, and county specific records. You will be able to determine if a company is registered as a corporation in the state, when they became incorporated, and if they are in good standing with the state. This site will also allow you to access property records in any given county. You will have access to the owner, value of the property, mortgage company, taxes on the property, and whether or not the taxes are currently paid. This is a good place to start if you are considering purchasing a piece of property.

Reverse Directories
www.555-1212.com provides cross-reference directories. You can find a person's address and telephone number by just having their name. A great feature about this site is you can do reverse searches. If you have a telephone number, you can get the name and address that it is listed to, or you can do the same with just an address. The site also has searches available in other countries. This site allows the user to find a person, or business, or to verify information.

Business Reports
www.D&B.com provides business reports on companies. This site will provide information on the owners of a company, their

industry, financial stability, average debt, and much more. The user can determine if a company appears to be financially secure before dealing with it.

Vital Records

www.vitalrec.com provides access to birth, death, and marriage records. The site can help an individual track down relatives and/or confirm the identity of someone who appears too good to be true!

Social Security Traces

www.iqdata.com provides a method of obtaining a person's social security number. If the user supplies the person's name and last known address, a social security number is provided. Once the number is obtained, a trace can be conducted through the same site to determine the person's current and previous addresses. Once again, the site will help locate someone that skipped out on a debt or confirm a person's identity.

You can see from these links that a wealth of valuable information is out there. The question is "why get scammed" when you can quickly check them out?

Mr. Kelly Riddle has more than 20 years of investigative experience. He earned a Bachelor of Science degree in Criminal Justice from the University of North Alabama and was chosen as the "Private Investigator of the Year" by the National Association of Investigative Specialists. PI Magazine named him as the "Number 1 PI in the United States" in its March 1998 issue. He is an expert in surveillance, insurance investigations, nursing home abuse, and computer investigations, and he has been chosen by The National Association of Investigative Specialists as "One of the Top 25 PI's of the 20th Century."

His prior law enforcement experience includes being a member of the SWAT team, a Training Officer, Emergency Medical Technician, Evidence Technician, Arson Investigator, Juvenile Specialist, and Traffic Investigator.

Mr. Riddle is the author of nine books, four videotapes, an Internet Investigation software package, and more than 40 articles. He has been the guest speaker at more than 250 events and has been on national TV and radio and has been featured in newspapers across the country. Mr. Riddle is also the Founder and President of the PI Institute of Education. Mr. Riddle is Founder and President of the PI Institute of Education, as well as of the Association of Christian Investigators with more than 400 members in 43 states and six countries.

Identity Theft Prevention
By Brian Keltner

I want to share with you some information that may keep you or someone you know from being another Identity Theft statistic.

We have all heard the horror stories about falling prey to one of these Identity Theft perpetrators.

The nightmare of identity theft occurs when someone gains access to another person's personal information, such as social security number, driver's license number, bank and credit card account numbers, and uses them to commit fraud or theft. An imposter can use your identity to open fraudulent credit accounts, secure loans for cars and housing, or steal money from your bank accounts.

I want to pass along a few tidbits on how to protect your information and what steps to take if you become a victim.

Home/Office
Shred EVERYTHING with a crosscut shredder. This refers to everything that contains your information on it, including any invitation for bankcards. Someone could use your invitation for their fraudulent use.

DO NOT set your outgoing mail in the mailbox for pickup. DO NOT put your outgoing mail in the Mail Box the night before pickup or when it does not get serviced often. Fraudsters can reach into the mailbox to steal mail for the purpose of obtaining your information and/or checks.

DO NOT give your information out to anyone over the phone if you did not initiate the call. Be cautious of giving your information out to a company with a toll-free number. Some companies use prison workers to handle their reservations and telemarketing efforts.

Shopping/Dining
Pay attention when using your bankcards. Keep track of your purchases and where you made them. Avoid letting your bankcards get out of sight, such as running a tab at a restaurant or bar. Be aware of employees skimming your card through anything other than the cash register. They may have a portable skimming machine (the size of a cell phone) that captures your card's strip information. The information that is contained on the magnetic strip of your cards is much more valuable to a fraudster than the card itself.

Internet
DO NOT e-mail any information that you do not want to make public. This includes your personal and bank information. Purchase only from reputable companies. Use abundant caution when making a purchase at an online auction, such as EBAY where the seller uses an anonymous e-mail address such as Hotmail/Yahoo.

Steps To Take If You Become a Victim of Identity Theft
1. Cancel your bankcards. Have the phone numbers and account numbers handy. You do not need the added stress of searching all over the place for these numbers.
2. Call and make a report with the law enforcement agency where the crime was committed, if known. If your wallet/purse is stolen while in New York, then contact the New York Police Department.
3. Call the credit reporting agencies to place a fraud alert on your account. This will help stop the perpetrator from opening other accounts under your name.
4. You may also contact the Social Security Administration Fraud Hotline and the Department of Motor Vehicles to report the theft of your information.

Telephone numbers that will be of help to you.

Equifax 1-800 525-6285
Experian 1-800-301-7195
Trans Union 1-800-680-7289
Social Security Administration fraud line 1-800-269-0271
U.S. Federal Trade Commission 1-877-438-4338
California DMV's number to report ID theft 1-866-658-5758

For more information on how you can protect yourself from fraud the U. S. Government's Identity Theft website is at www.consumer.gov/idtheft/.

Identity Theft Check List

CANCEL CREDIT CARDS

Bank Name   Account Number   Phone Number

_____

_____

_____

_____

_____

_____

MAKE A REPORT WITH LOCAL LAW ENFORCEMENT

Agency Name _____

Phone Number (___)_____

Report Number _____

Date/Time _____

CALL THE NATIONAL CREDIT REPORTING AGENCYS

Equifax 1-800-525-6285
Experian 1-888 397 3742
Trans Union 1-800-680-7289

ADDITIONAL CONTACTS

U.S. Federal Trade Commission 1-877-438-4338
DMV Fraud Line In Your State . . . . _____
Social Security Admin. Fraud Line 1-800-269-0271

Use the back of this form for additional information, i.e. driver's license number, gas cards, department store cards, calling cards, etc.

Reducing Shrinkage
By Roger H. Schmedlen

Shrinkage is a stylish word for theft. In convenience stores and similar small retail operations the primary shrinkage is generally attributed to shoplifting and employee dishonesty, with limited losses from holdups and burglaries.

To reduce exposure from shoplifters, the franchise owners should adopt the goal of preventing or minimizing losses, as opposed to catching thieves. The owner's objective in business is to earn maximum profits. It is a more economical and efficient policy to deter a potential shoplifter than to apprehend and prosecute one caught in the act.

A logical physical layout of the store is the first step in deterring. Stores that place small desirable items of relatively high cost near the entrance normally suffer excessive losses. Low cost bulk food items such as bread are often seen in reasonably secure areas, while displays of film, sunglasses, lipstick, makeup and other easily concealable items are place in vulnerable areas by the entrance. Stores such as these can reduce exposure by relocating merchandise according to priority. Primary shrinkage items (determined by past inventories) give the owner a guide to designating priorities.

Employee awareness is another deterrent. Conscientious personnel who keep a close eye on suspicious persons and prohibit loitering in the store tend to encourage the potential thief to seek easier hunting grounds.

Statistically, over 70% of shoplifters apprehended during one recent year were operating between noon and 9:00 P.M. Thus, it makes sense to schedule the most conscientious clerks to work during this period.

Again statistically, shoplifting occurs at approximately the same rate each day of the week, and December is the only month where it appears convenience stores suffer appreciably increased losses. The twelve to seventeen-year-old age group is the most active shoplifting group in relationship to its size, followed by the eighteen to twenty-nine age groups. Persons over fifty years of age are least likely to shoplift.

Security guards at a convenience store are generally unwarranted and cost-prohibitive, but they may be necessary in some areas. In some situations armed guards may be hired to cover parking lots. However, under no circumstances should an armed guard ever work inside a convenience store, or for that matter, any retail store. Armed guards in retail stores are not assets, but are serious liabilities.

If guards are required, the convenience store owner should not look for a bargain rate. The owner should seek a quality service that adequately trains its guards and pays those assigned to the store a reasonable wage. Obviously, the rate to the storeowner will be higher than from a service providing untrained, minimum wage watchmen, or those who may be taking advantage of a government subsidy program.

Prior to contracting with a guard service, the franchise owner should be provided with a proof of the contract service's liability insurance. One million dollars liability coverage (with errors and omissions and other applicable riders) from a solvent carrier and statutory workers compensation coverage should be considered a minimum requirement, and deductibles should be reasonable. An additional insured endorsement is often available, providing some measure of protection to the storeowner in the event of a lawsuit.

Should shrinkage warrant consideration of physical security controls or systems, the most effective and economical approach appears to be a video system placed where it is obvious to customers and advertised at the entrance and other strategic locations. "Fake" or replica cameras are often quickly exposed as being spurious. Additionally, the presence of such replicas can significantly increase liability should a theft occur on the premises. Although clandestine CCTV systems provide no ongoing deterrent since their presence is unknown, they can prove useful when employees are suspected of theft or fraud. Obviously, video cameras, either overt or covert, should never be installed in areas where customers or employees have a reasonable expectation of privacy.

A quality CCTV system including a videotape recorder and monitors, at least one of which is visible to the public, is a proven deterrent with related benefits. When thefts do occur, prosecution is simplified. One convenience store franchise holder, for instance, recently reported a number of videotapes being held by police as evidence in shoplifting cases.

Related benefits of the CCTV system include protection from employee theft, which in the retail environment accounts for the greatest amount of shrinkage. A decade ago the National Institute of Justice estimated the U. S. retailer suffered losses in excess of $4 billion annually through employee thefts.

According to the National Retail Merchants' Association, an effective CCTV system can cut theft losses by 75% in as short a period as six months. However, this reduction will only continue if employees believe that tapes are being viewed or at least spot-checked by managers or owners.

A well advertised visible CCTV system can also act as a holdup deterrent. One franchise owner stated at a recent convenience store association meeting, "Had it not been for the (CCTV) system, I might not be here today." He explained that when three armed men entered his store intent on robbery, they panicked and ran out as soon as they observed the video camera.

After another franchise convenience store owner's life was threatened at his store, the police picked up the guilty party and the problem was resolved due to the video system at the store.

In relationship to its effectiveness, CCTV has to be considered one of the greatest bargains of our times. Although there are firms that offer leases on such systems, in recent years the cost of high quality CCTV systems has dropped to a point where convenience stores and other small retail businesses can obtain an excellent return on investment by purchasing these systems outright.

Roger H. Schmedlen has over 30 years investigative and security consulting experience and has handled projects in Australia, Europe, Southeast Asia and Latin America as well as throughout North America. He has acted as an expert witness on approximately 50 lawsuits involving security/loss control negligence.

Interview or Interrogate
By Edmund J. Pankau

What is the difference between an interview and an interrogation? Most of us see an interview as something that is done in a pleasant surrounding by people with smiling faces who are sharing information in a non-confrontational manner. When we hear the term interrogation, we think of bright lights, rubber hoses, and the SS officer in the black leather trench coat uttering that famous line, *"Ve have vays of making you talk."*

The rule of thumb that is that an interview is something you do to third parties having knowledge of a subject, and an interrogation is reserved for the subject of that investigation. The interviews are used to gather information about the subject and to ask those probing questions that make the interrogation successful if the subject is showing signs of deception.

One of the key elements in conducting an internal investigation, or any investigation in which you are trying to determine if one of a number of people are doing something wrong, is to develop the information already in the minds of their co-workers, the people they work with everyday.

The people having the most direct and detailed knowledge of the daily activities of anyone are the employees and personal friends that see the subject on a regular basis. Our job is then to determine the opinions and knowledge of these employees by asking a series of questions designed to get their suspicions, intuition, and gut feelings in a non-intrusive, non-combative form through a series of questions that tell us what they know, without them knowing why we want it.

To overcome the reluctance on the part of many people to discuss their suspicions about other people or to specifically name an individual that they believe is involved in wrong doing, we have developed an interviewing technique that identifies those parties that fall outside the parameters of accepted peer behavior. It helps identify those people whom the interviewed party believes may be involved in or are responsible for the problems we are investigating.

The key to this interviewing technique, which I call the "queen for a day," is a peer-rating system in which the interviewed party appoints a numerical value to each person whom they work with or have knowledge of in their daily activities. These interviewed parties then rate those individuals on a point scale of one to ten based on their general, overall, nonspecific perceptions, impresssions, and gut feelings that tell us whom they suspect of being the cause of our client's problems.

This queen-for-a-day interviewing and rating system allows workers to rate the other people they know or work with, and identify those individuals they consider outside the norm or standard. It allows the investigator to determine whom the majority of employees or other people rate most poorly. Through this interview technique and its rating system, people identify the individuals they feel are responsible for the problems that they perceive.

This queen-for-a-day technique is so named because the interviewer tells the employee or knowledgeable party that, for the purposes of this interview, the interviewed party has just been promoted to the position of supervisor, manager, or president of the company. The "queen" is to rate every other person in the company based on the personal perceptions and impressions of each employee. The interviewed party assigns a one as a lowest rating and a ten as the highest rating to the other parties based on their impressions and preferences.

This interviewing technique is a useful tool that helps to focus the initial investigation on parties that the majority of other people believe fall outside the accepted behavior pattern, so that investigators can then concentrate their investigation on those whom the majority suspect. If the majority of co-workers rate an individual lowest in the group, then this tells the investigator that the majority feel that some activity has or may have taken place, that they disapprove of this individual in some way, or that they have knowledge that this person is really a scum bag.

After interviewing all of the employees or potential parties, the investigator determines the party or parties who scored the lowest on this peer rating system and then examines the ratings assigned by those low-scoring parties in comparison to the others in the test program. If these lowest-rated individuals rate other employees in a range contrary to the majority, the investigator then knows those parties whom the lowest-rated individual (and highest potential suspect) feels closest to and most comfortable with. These individuals are often found to be closely related to the activities or dealings of this prime suspect. These findings also indicate that the person scored most highly by the lowest-rated person and also rated lowly by other peers are those often involved as an assistant or co-conspirator with the lowest-rated suspect.

Once the abnormal or lowest-rated persons are identified through this interviewing technique, then further interviews of a more in-depth and probing nature can be used to determine the nature and extent of the concerns or problems known by the interviewed parties and to determine which employees have the most knowledge of problem areas or would be most cooperative in helping in the investigations of the subjects. This technique is very simple to use and master.

Once understood and practiced, this method can be an extremely effective tool for any investigator, auditor, or fraud examiner and is an excellent training tool for developing higher levels of investigative interviewing techniques.

The rules of this test are as follows:

1. Score people only by number 1-10, with 1 being super bad and 10 being super good. Base it on gut feelings, intuition, hunches, or something one can't explain.
2. Explain, if asked, that you are evaluating negatives such as lies, exaggerations, bragging, thefts, insincerity, and just being sneaky.
3. All numbers are kept confidential.

I have used this technique for more than 12 years and have shared it with hundreds of other investigators around the country. One of them recently used this simple technique in the investigation of drug losses from a hospital pharmacy, and in her first tests she identified the ultimate person found to be responsible for the thefts.

If you don't believe me, just try it yourself. You will find that this is the single best interviewing technique when fishing for information. It will point to the problem people and problem areas almost every time.

Edmund J. Pankau, President of Pankau Consulting, is one of the world's leading authorities on the detection and prevention of financial fraud and on the conduct of domestic and international financial investigation. During his thirty years of investigative work he has assisted in the recovery of millions of dollars and in the prosecution of criminals throughout the world.

He is an internationally recognized author and trainer who conducts seminars on the prevention and detection of fraud, hidden asset location, the conduct of due diligence examination, and the protection of personal privacy. He is also the author of numerous articles published in professional and business journals and of several best selling books, including *Check It Out, How to*

Make $100,000 a Year as a Private Investigator, and *Hide Your Assets and Disappear*, which was a New York Times bestseller.

Mr. Pankau has been featured in USA Today, Time, Business Week, Wall Street Journal, and New York Times and has appeared on ABC's 20/20, MacNeil-Lehrer, BBC-TV London, CNN-Money Line, Inside Edition, Larry King Live, America's Most Wanted, and many other radio and television programs. A graduate of Florida State University's Department of Criminology, he is a Diplomat of the American Board of Forensic Examiners. P.I. Magazine named him one of the world's top ten investigators.

The Enemy Within
By Scott Castleman

When it comes to crime, employee theft and fraud are as American as Apple pie. Today's business and industry face an uphill challenge from the moment they open for business. Some areas of theft are easier to detect than others while white-collar fraud can take years to discover. Any way you look at it American business owners lose billions annually to crime. We would like to think that the main perpetrator of crimes would be the shifty criminal, labeled as a "menace to society" that wears a ski mask and robs business owners at gunpoint. Unfortunately, although this does occur, 35% of all business failures are attributed to employee theft and fraud. Although there are literally thousands of ways a business can become victim to their employees, we will examine two of the most common areas—employee theft and fraud.

Employee theft is defined as theft of profits from a business by employees. Every year millions of employees steal billions of dollars worth of merchandise and cash from their employers. Current rough estimates reveal that dishonest employees annually cost businesses and industry twice as much in cash and merchandise as do all the nation's burglaries, car thefts, and bank robberies combined. These astonishing acts are not being committed by "street criminals" whose immoral behaviors are presented daily by the media as a "menace to society." They consist predominately of solid, well-educated, respectable citizens who pay taxes and believe firmly in setting good family values, as well as in the virtues of hard work and honesty. These individuals are also quickly infuriated by the thought of welfare abuse, street thugs, and the literalness of immorality, which they believe is upsetting the established social order. Ironically, these individuals do not regard themselves as criminals.

All instances of internal theft are not the same. Employees have numerous reasons and justifications for embezzling money and stealing merchandise, which is typical in the retail trade. A vast amount of internal theft is committed indirectly by such means as discounts, loans, and acquiring damaged goods.

I was recently speaking with a client who advised me that he has never had any internal theft issues, but shoplifting was a huge concern. I then told my client, "How fortunate you must feel, having the only business in the United States never to have had a dishonest employee." It is this sort of naïve thinking that forces many retailers out of business. In fact, a major insurance company recently reported that one in three small business failures can be attributed to internal theft. Let's face it, dishonest employees steal more before lunch than shoplifters steal all day. It is genuinely naïve to believe that employees, whom you trust implicitly, will not betray you. Not one single employee is above reproach, however difficult this is for you to believe. Internal theft is a time-honored profession in which its perpetrators know exactly what they can and cannot get away with. The easiest way to embezzle money from an employer is to be well liked, perform one's job well, and appear to be above reproach.

Most employees do not start out as thieves. Several symptoms must present themselves in order for an employee to turn dishonest. A symptom of what? Well, a symptom of what the employee perceives as:

- Inadequate pay
- Not receiving promised raises
- Pressure to achieve unreasonable goals
- Unfair treatment by management
- Overworked and under appreciated
- Favoritism
- Lax methods of discipline
- Unwillingness of management to listen

To many employers these symptoms may be real or imagined, but to the employee they are real and at this point many employees begin looking for ways to retaliate. In all incidents of theft a triangle is present, known to professionals as the theft triangle. It consists of three elements:

(1) *need or greed*
(2) *rationalization*
(3) *opportunity.*

Once the rationale to justify embezzlement is established, all that's needed is the opportunity. Employees generally will not steal from an employer whom they look up to and respect, one who appreciates their efforts, who treats them like human beings instead of puppets, and who sets an example of impeccable integrity. However, they will steal from an employer who allows their frustration and resentment to build up to a boiling point. Commonly, embezzlement is simply a form of retaliation. In order to combat and reduce employee theft you must eliminate the "opportunity" by establishing a solid written loss prevention program. By doing this you will greatly reduce your risk of theft incidents by your employees.

It is important while establishing these programs to keep in mind the following statistics: 10% of your employees are honest and will remain honest no matter what polices are established, 10% will steal regardless, and 80% are generally honest and will remain honest if you create an environment which discourages and detects theft. As the risk of getting caught increases, the probability of theft decreases.

Detecting dishonesty within the workplace is not always a simple task. However, looking for signs or red flags is the first step to prevention. Red flags to employee dishonesty can either be found in an individual's behavior or through company records. An alert business owner or manager should look for the following signs and should consider further investigation when any of them appear:

Red Flags To Internal Dishonesty
- Frequent visits by friends
- Excessive use of alcohol or drugs
- An abrupt change in lifestyle (*living beyond their means*)
- Close social relationship with suppliers and or customers
- Frequent borrowing of money from other employees
- Obsessive neatness in work orders and bookkeeping
- Refusal to take vacations
- Taking long lunch breaks or frequently arriving or leaving late

Confession of a Teenage Employee
Told by an eighteen-year-old record store clerk

I got this job at a local record store. It was a big national chain and they made lots of money. I got the job because I was in a local band, sort of the NW grunge thing and I knew a lot about music. It was no secret that all the managers of the store were drunks. So most of the time the employees took advantage of that weakness. Everyone in the store knew they left for lunch at 12:00 and would not return for the rest of the day. It started for me by taking a few bucks out of the till every day for lunch and smokes. An employee taught me how easy it was to make change to customers without even ringing in the sale. The only hard part was remembering how much money to take out. The more I took the harder it was to stop. It almost became addictive...like a drug, I had to do it.

Eventually I didn't even look forward to picking up my check on payday because it paled in comparison to what I made out of the till weekly. The funny thing is I was not the only person doing it; just about everyone had his or her hand in the till. I think even the managers were sneaking a few bucks. Sometimes to make a little extra money I would stick a bunch of CD's in the trash and pick them up later from the outside trash and sell them to other record stores.

  Yeah, they had undercover security but everyone knew they only made $5.00 an hour, so they were no real threat. It kind of made it more challenging. I guess I really did not need to steal money because I lived at home and had no bills. But it was just too easy. And management was just so stupid. I no longer work at that record store. I got a job that I actually enjoy now and the pay is a little better. —Drew

Prevention Stratagies

Conduct Background Checks. Any position you fill within your business should be begun with a complete pre-employment background check. Retain a reliable firm to perform the inquiries. No exceptions should be made.

Set a Policy. Establish a strict internal theft policy and stick with it. It is time to move beyond minor punishment or termination and prosecute your dishonest employees. Your employees must realize that you will not tolerate theft.

Establish Internal Controls. When a lack of internal controls is present, so is the opportunity to steal. Eliminate the opportunity by establishing standardized controls, which provide consistency and uniformity that will reduce incidents of fraud.

Educate Employees. Show your employees how bottom-line repercussions of theft affect the company and encourage their help. Outline procedures for reporting theft, ensure confidentiality, and lay out the consequences of dishonesty. Clear and concise policies provided in writing at the time of hire must be continually reinforced, and continuing education to all employees is critical to how seriously your employees will take your policies.

Use Covert Investigative Techniques. There is nothing more harmful to employee morale, productivity, and profits than upsetting loyalty. It is easy to create a climate of suspicion and mistrust by clamping down in highly visible ways that convey a "we don't trust you" image. An effective program must be capable of prevention, yet be aesthetically pleasing.

Remember, when employees steal from their employers, the underlying factor is most often some form of injustice, either real or imaginary. An alert business owner works diligently to avoid both the reality and the perception. If you treat your employees like members of a team, they will reward you with honesty and loyalty. Treat them like puppets and they will constantly be looking for ways to outsmart the puppeteer. Most business owners, because they are only human, become a part of the problem. Even after a theft occurs and the employee is identified, the manager typically tends to think of all the "good" things that employee has done and exhibits compassion for the employee.

Acceptance of any form of dishonesty, for any reason, is a mistake. This misled compassion becomes the very base, the underlying cause for most of the corruption it seeks to correct. Compromise has replaced the moral absolutes that have held our society together. And such compromise is the singular reason our society is falling to pieces. Shoplifting, fraud employee theft and disregard for the law are not the problem. They are just a symptom of a generation, which does not stand for anything.

Corporate Fraud

Corporate fraud is defined as a crime committed by an individual in a company with or without the company's knowledge in an attempt to increase profits. These criminals are not your ordinary street thugs. Most are well educated, intelligent, and hard working businessmen and woman who have fallen prey to the criminal element and have seeped into every nook and cranny of American society. White-collar crime and fraud are very difficult topics to discuss, due to the large variety of crimes committed within this category. It runs the gamut of:

- Large-scale embezzlement (from 1950-1971 100 banks failed as a result of employee embezzlement)
- Bankruptcy fraud (illegally accepting bankruptcy to avoid debt)
- Insurance fraud (fraudulent insurance claims or intentional destruction of property for purposes of collecting insurance money)
- Medical fraud (doctors and physicians and their staffs taking advantage of medical benefits programs)
- Dangerous foods (e.g., Hormel Foods reprocessing spoiled meats)
- Deceptive advertising (using advertising to mislead custommers about products)
- Price fixing (businesses conspiring to keep their prices artificially high)
- Illegal or unsafe work conditions (Union Carbide was cited 266 times by OSHA for not providing employees with safe breathing devices)
- Theft of public resources
- Consumer fraud
- Corporate espionage

Although the above examples do not even begin to scratch the surface of the problem, it allows you to understand the importance of identifying and protecting your company from such fraud and deceit. Often high-ranking management personnel are the ones who commit the crimes. This type of crime costs Americans in excess of $500 billion. As far as illegal or unsafe work environments, the loss of life is estimated at approximately 100,000 deaths per year. Homicide and manslaughter are estimated at approximately 20,000 annually.

Top Corporate Criminals:

Exxon Corporation	Crime: Environmental	Fine: $125 million
Bankers Trust	Crime: Financial	Fine: $60 million
Sears Bankruptcy Recvry. Mgt.	Crime: Financial	Fine: $60 million
Damon Clinical Laboratories	Crime: Fraud	Fine: $35.2 million
Pfizer Inc.	Crime: Antitrust	Fine: $20 million
Royal Caribbean Cruises Ltd.	Crime: Environmental	Fine: $18 million
Northrop	Crime: False statements	Fine: $17 million

Fraud can be for the benefit and gain of an individual, a business, or an industry. When an individual commits fraud the benefits may be direct, such as receipt of money or property, or indirect such as promotions or bonuses. When a business commits fraud, the benefits to the business are usually direct in the form of financial gain. Some states have specific laws relevant to fraud. Other states have laws that deal with the fraud on a more specific level, such as larceny, bribery, embezzlement, or some other specific crime. Most business owners have a difficult time discovering that fraud has occurred until it is too late. It is essential that business owners maintain strict internal audit policies and adhere to them. If something does not seem right, contact a private investigation firm, which specializes in corporate fraud. Investigation firms with detectives who are Certified Fraud Examiners will probably be the most suitable. Business owners have an obligation to be alert to the possibility of crime within their business or industry since not being aware could result in the loss of life or of millions of dollars.

Tracking the Global Criminal
By Edmund J. Pankau

Catching today's sophisticated white-collar criminal is like peeling an onion—going through layers and layers of business entities, paper trails and false identities, ring by ring, coming ever closer to the core of the crime.

If you have read any of the recent stories about Marty Frankel, a one time stockbroker and current white-collar con poster boy, you would see that he fits the perfect profile of this kind of criminal. Marty conducted business with many false identities, and to improve on his chances of disguising his activities, hid behind numerous corporate shell businesses and trusts so that gullible investors like trust and insurance companies would give him millions of their dollars to invest.

Once investors started to question the bounced checks, unpaid dividends, and cooked books, Marty fled out of his three-million-dollar home and expensive toys and flew to Europe to play the game of hide and seek. On the way he bought ten million dollars worth of diamonds and a few gold bars to have pocket money, just in case.

Not too many years ago Marty may have gotten away with his millions, but today's computer technology and online databases, which provide instant access to information around the world, pinpointed every move Marty made. The FBI quickly identified Marty's parents, business associates, and investors, as well as his many girlfriends drawn to his money like bees to honey.

By following up on leads to his European business associates and flight plans filed on his chartered jet, federal agents targeted Marty's activities in Rome and trailed him to Germany where he was arrested and charged with money laundering and fraud.

To have even a hope of getting away, the modern white-collar criminal has to make complex plans to "muddy the waters" in order to hide his financial trail. He does this through a web of complex maneuvers designed to disguise his true identity, intent, and the ultimate destination of his newfound wealth.

When conducting an investigation into the activities, business dealings, and assets of a person doing business in foreign countries or hiding assets outside the United States, an investigator must gather pertinent information from a variety of sources and seek to piece together the web created by his target.

To conduct a successful foreign asset investigation, the prudent attorney or investigator must document any evidence of the following facts or issues:

- Document the foreign travel of the subject to a specific country
- Determine the subject's business and financial activities in the foreign country
- Discover the location of bank and brokerage accounts and the subject's business relationships to that country
- Locate and interview the parties having access to such information

A number of resources are available to help investigators determine if an individual has traveled to or through a foreign country. The most widely known resources available to state and federal police agencies, is a database developed and maintained by U.S. Customs Service, now known as Treasury Enforcement Computer Service (TECS). This database, formerly designated as EPIC (El Paso Information Center) records the reentry of each person into the United States from a foreign port using the information that each individual fills out on a U.S. Customs form that documents the reentry point and goods purchased and declared. This database is commonly used to document the frequent foreign travel of parties believed to be involved in drug smuggling, money laundering, or other types of crimes investigated by the Customs Service.

The information in TECS is reported to the Financial Crimes Enforcement Network (FinCEN) in Washington, D.C. FinCEN then provides the information to law enforcement and regulatory agencies interested in determining the travel history and activity of parties under investigation by state and federal agencies investigating drug, financial, regulatory, or organized crime matters.

On an international level, INTERPOL has the ability to locate and document foreign travel between most countries and will often visit the offices of the International Air Transport Association (IATA) to review their records that document airline travel on the international level.

A thorough investigation can disclose a number of other methods to determine if a person has traveled to a foreign country. Some of these methods most often found to be successful include the following approaches:

When a suspect is arrested for any reason, a thorough physical search should be conducted to determine if the subject of the investigation (or his or her traveling companion) is carrying a passport or documents indicating foreign travel or business activities. If such records are found, an investigator should copy the pages recording foreign passport stamps and visas. The investigator should also look for hotel receipts and amenity packages, as well as matchbooks that identify the locations that the suspect stayed at while visiting these countries.

One of the easiest ways to determine the foreign business activities of an individual is to examine his or her business, personal, and mobile telephone records to identify countries, cities and parties called. These records will often lead to a bank, attorney, business agent, or broker that the subject has contacted to set up a business or financial relationship in a foreign country and may lead to a bank account, safe deposit box, or assets that would otherwise never show up in their name or control.

In either a criminal or civil court case telephone records can be subpoenaed by the prosecutor or plaintiff attorney's to establish and document the telephone toll record of the person being investigated and to learn about relationships with other individuals who might not be otherwise contacting them in any other way.

An individual's credit card statements are another valuable source of information for determining foreign travel locations and expenditures. These records often reveal a detailed account of airline tickets, hotel accommodations, cash withdrawals, and purchases made by the subject in the foreign country. These records may also detail the time line of the target's foreign travel and place them in a location at a time important to your case.

In one case, investigators traced an economic ministry attaché from Colombia, a man suspected of embezzling $12 million, by tracking his American Express charges from Colombia to the United States and then to Israel and Austria. Working with the credit card security department, credit card investigators were able to document the daily financial business and travel activities of the attaché and track his movement through hotel reservations and restaurant charges made to his American Express account. This information provided a detailed history of the attaché's activities during the period of time crucial to the investigation and helped place the attaché in locations where the bank accounts were located.

While checking the record of bank transactions, you or your investigator should also look for information concerning the purchase of traveler's checks, money orders, wire transfers, or overnight packages. Be aware that the messenger services can track a delivery history for six months to a year, incoming and outgoing, just from the bar code on packages. With this information the astute investigator can determine where checks, money orders, or traveler's checks were cashed, which well may be the same location or bank where the subject opened an account. Today you can't cash a check unless you already have an account in that bank.

The ultimate destination of wire transfers, telexes, or overnight packages may also be the financial institution, attorney, or agent with whom the subject set up foreign business activities and accounts, or may even be the mail drop by the offshore enterprise.

If you are unable to locate foreign accounts by following the money, the next step is to locate people who know the subject's personal and business activities, including enemies, friends, and former employees. If someone likes you, they may tell three or four people. If they hate you, they will tell EVERYBODY. An executive's travel plans are usually made through a secretary and travel agent. Both of these sources may be able to contribute reliable information to an investigator who looks for them or an attorney who subpoenas these witnesses and their records. Another possible resource is a traveling companion (a la Marty Franklin) who accompanies the target on a little vacation or trip, particularly when the destination is Bermuda, the Bahamas, the Cayman Islands, Belize, or another similar tropical financial paradise that asks no questions and does not tell the FBI.

In one case, a secretary informed investigators that her former employer was hiding money overseas by sending it out of the country with his attorney. The secretary gave specific details of the dates and times of the foreign travel, and the information was transmitted to U.S. Customs, who gladly saw that the attorney was arrested when he attempted to leave the United States without declaring to customs that he was carrying cash in excess of $10,000, as required by law.

Once the attorney was searched and the money confiscated, he realized he had a choice: he could either state that the money was his own and face income tax evasion charges from unreported income or admit that he was acting as an agent for another party and face the lesser transporting charge. This transportation charge would probably be dismissed if he cooperated in the bigger case against his client (it's the IRS godfather offer, the one you shouldn't refuse). Which do you think that he chose?

Finding the target's former secretary or personal assistant is also crucial to the financial and business aspects of the case, particularly on the issue of the intent of the parties. The secretary is most likely the person who notarized the subject's business documents, filed his papers, and contacted his clients. To find this little gem, look at other financial documents executed by the subject (to find her name or initials on letters) and then call the state notary board to determine the current registered address and bonding agent of the secretary or notary who worked for your target.

Since they are usually responsible for filing, recording and notarizing company records, corporate secretaries often know of and feel responsible for improper, immoral, and illegal activities conducted by their employers. Sometimes these employees even seek to protect themselves from future litigation by keeping personal diaries or retaining copies of documents they think may compromise their position if their employer's scheme is discovered and THEY are visited one dark night by Uncle Sam. (Remember that old line, "I'm from the Government and I'm here to HELP you"?).

Since a secretary often signs checks for her or his employer, they can, according to IRS and tax court decisions be held liable for the illegal activities perpetrated through the use of those checks. Any smart secretary feels either a natural hesitancy to sign such documents, or when forced to do so, keeps her own records of the circumstances surrounding the events to protect her in case of future problems.

In the course of conducting an offshore investigation, investigators should question the target's secretary or personal assistant and determine if he or she has the following items or information:

- A personal diary that details the employer's travel and business activities
- A personal Rolodex of numbers frequently called for the employer
- A notary log that records documents that the secretary notarized
- Telephone directories the secretary kept to remember the employer's key contacts

- Copies of documents made because of the secretary's personal concerns about the employer's ethics or business activities

To Locate A Target's Travel Agent

First try his or her Rolodex file if it is available. Business people usually have a favorite travel agent and stay with them for years, although with new Internet sites like www.lowestfare.com, many travel agents have bit the dust. If there is a travel agency in or near the office building of your target, give that agency a call to see if your subject or his or her company had an account with them.

If the target did have such an account, the agency should have at least a three-year history of the subject's travel and can tell you exactly where and how often he or she traveled and where they stayed, if the agency booked the hotel as well.

If you can't find the travel agent, consider a subpoena for the records of the Airline Reporting Corporation (ARC) in Washington, DC, the clearinghouse for airline tickets of all the major air carriers. ARC compiles and collates all domestic air tickets issued by domestic carriers. The sister organization, the International Air Transport Association (IATA), headquartered in Montreal, collates the records of foreign travel for many of the world's airlines. The records of ARC and IATA can help an investigator determine the name and location of the travel agency used by the subject. The agency can then review where and when the subject traveled.

Many offshore investigations begin with information provided by fired secretaries, spurned spouses, or disgruntled employees who harbor a grudge against the wealth or illegal activities of someone they know or have had some type of relationship with. The records they keep often provide the missing link necessary to prove the existence of foreign banking relationships or hidden business interests and relationships.

Consider this case, for example:

Sheila was the faithful secretary and mistress of one of the nation's largest pornography dealers for more than 12 years. As his confidant and bookkeeper, she traveled with him on Caribbean vacations and accompanied him when he set up several Cayman Island bank accounts, where she observed that he deposited $5 million during the course of her employment.

As she grew older and lost her physical charms, Sheila's employer became obsessed with younger women. He started bringing younger women into the organization, women who ultimately took over Sheila's relationship with him as well as her secretarial duties. He did not consider Sheila a threat and ultimately fired her from her job.

Forced into the job market and unable to find a well-paying job, Sheila became disillusioned and bitter over her loss of status and position. She became a willing witness when an IRS agent investigating tax matters in the pornography business approached her.

Sheila provided oral testimony about the trips to the Cayman Islands and the existence of Cayman Island bank accounts in the pornographer's mother's maiden name. She produced copies of deposits and overnight package receipts documenting deposits made to the Cayman Island accounts set up to hide the money made from pornography and drugs by her former employer. With this evidence, the government agent was able to document no only income tax fraud, but also income from drug dealings, which gave the US attorney sufficient grounds to obtain an indictment and conviction against the pornographer and his business.

As a result of his poor employment decisions, Mr. Pornography is now a guest at Club Fed, where he will probably be for an extended vacation.

Once an individual's foreign travel has been determined, the next step is to find out where the subject stayed in the foreign country and what type of business activities were conducted there.

One of the best sources for this information is the credit card and travelers check activity of the subject. The ease and convenience of credit cards work to the advantage of the investigator and the disadvantage of the target individual. Because of the ease of their usage, most people who travel overseas charge their airfare and hotel bills on credit cards so they don't have to purchase or exchange foreign currency. (Just try to buy an airline ticket, rent a car, or a hotel without a credit card.)

Major credit cards are quick and convenient to use because they accept all major currencies, but this efficiency creates a paper trail that documents the target's itinerary and activities. By identifying information about the hotel used by the subject, an investigator can locate, document, and examine the hotel's telephone records to find out whom the subject contacted during his or her stay, and therefore, where they conducted their business.

Examining a customer's hotel telephone record helped determine the existence of foreign bank accounts and the names of business agents used by an executive alleged to have transferred millions of dollars from a Texas savings and loan. Through the examination of credit card records, investigators determined the executive stayed at a hotel in Berne, Switzerland, and while at the hotel made telephone calls to two banks, an attorney, and a foreign private postal service located in that city.

Through a review of the daily photographs of the people entering the Swiss banks called by the subject (European banks photograph every person who enters their premises because of the much higher incidence of bank robberies in their countries), investigators were able to document that the executive visited a certain bank on a specific date. They found that the subject opened accounts with the proceeds transferred from his savings and loan into that bank and conducted a series of financial transactions that ultimately led to the bank.

The phone calls made during these hotel visits were crucial in establishing a link between the executive and a Swiss attorney with whom the executive did business. The attorney made purchases of real property in the United States for the benefit of the executive, who later claimed he was only the caretaker of these properties, which helped establish a link between the two.

By proving the existence of prior conversations between the executive and the attorney, the government was able to establish an ongoing relationship between the two that gave the court reasonable grounds to seize the properties purchased for the executive's benefit in the United States and prove a conspiracy between the parties.

In addition to examining the credit card statement of individuals traveling abroad, attorneys and investigators should look into the purchase of cashier's checks, money orders, or traveler's checks, negotiable instruments that are also frequently used to launder money.

Examining such financial instruments may also indicate the location of new bank accounts, the leasing of safety deposit boxes, and the purchase of extravagant gifts. In several cases investigators have even traced purchases of expensive foreign automobiles through the purchase of traveler's checks bought by the subject in the United States and cashed in Europe at the headquarters of Mercedes Benz.

An important step in an investigation of this type is to determine if the target individual has traveled overseas personally or if he or she has sent another party, such as an accountant, attorney, friend, or family member, out of the country with the assets they are attempting to hide.

An accountant representing a number of doctors in a suburban metropolitan area conceived of a scheme to take the unreported income made by his clients to the Cayman Islands where it was deposited in numbered accounts in the doctors' names. For years the accountant made a trip every month to the Cayman Islands, complete with diving gear and scuba tanks. Each month for six years the accountant illegally carried millions of dollars in scuba tanks past U.S. Customs and Cayman Islands Immigration each. This crime went unreported for years until the accountant's wife filed for divorce and turned her husband in for a reward on an income tax evasion scheme.

The IRS alerted U.S. Customs, which then monitored the flight plans of aircraft traveling to the Cayman Islands during the accountant's normal travel times. On his next flight the accountant was detained and the scuba tanks inspected. When customs' agents found more than $100,000 in cash inside envelopes inside the tanks, the accountant was charged with carrying more than $10,000 outside the United States without a customs declaration and a criminal charge was made by the IRS relating to unreported income of the monies found.

Facing the two charges, the accountant elected to plead guilty to the lesser charge of transporting money and gave the IRS an affidavit and testimony revealing the identity of the parties who had given him money to deposit in the Cayman Islands.

Sometimes the best evidence in the investigation of offshore matters comes from getting down and dirty, that is, bagging the trash. Like everyone else in the world, crooks generate garbage and often pearls can be found in the garbage can. Some of the best evidence is often a handwritten note or some personal scribbles made late at night and thrown away in the trash.

This evidence could include such pearls as envelopes that list a return address from parties having knowledge of the suspect, the addresses of foreign banks, confirmations from overnight packages sent overseas, or bank statements and checks that show the routing numbers and account codes where checks were cashed or funds were transferred through the target's bank to an offshore entity.

Most people throw away notes scribbled in haste while picking up messages from their telephone recorder. By conducting trash pulls on the subject's garbage, particularly at the end of the month, an investigator may find documents that inadvertently lead to key offshore business activities.

Certain days of the year, particularly holidays, are magic moments for any astute investigator. That's when individuals often call home to family and loved ones or send cards or notes showing their love and affection for their families. These magic moment holidays are as follows:

- Valentine's Day: Bag someone's trash after Valentine's Day and you may find return address from envelopes received from their lover or special friend. This is also a day to check for receipts for flowers, perfume, or candy on the charge card.
- Mother's Day: Even the most hardened criminals may call their mothers to let them know they are still alive. This is a good day to check the mother's phone bill to find a son or daughter hiding from the law.
- Father's Day: Not as many criminals call home to dad, but it is still a smart day to check the trash and phone bills when looking for errant sons or daughters that can't be found any other way.
- Thanksgiving: As a time of family gatherings, an investigator should keep a subject's family under surveillance to see if he or she goes home for dinner.
- Christmas: This is the time of year to check credit card receipts for purchases. Christmas is also the time when cards and envelopes are found from people who only make contact once a year.

In addition to addresses listed on the mailings, investigators can often find other treasures on the paperwork, such as fingerprints or DNA samples if they licked the envelopes, all of which helps establish the identity of the parties in a case.

No one single path or miracle method can make every offshore investigation successful. Hitting all of the bases, chasing down every possible lead, and finding where the suspect stumbled and left an opening into his or her financial affairs solve cases of this type.

The resources discussed here, if diligently pursued, will lead to the discovery of whatever evidence exists in many of the most complex cases and help show the true intent of the actions of the participating parties.

Each and every investigation teaches investigators new methods used by targets to hide money, and avoid and evade the law and the court system. An investigator's job is to catch those mistakes made in haste and put together the puzzle that the subjects so carefully built and then destroyed hoping no one could find and put all the little pieces back together again.

Developing an Anti-Fraud Program
By Jeffrey A. Russell

With the rapid spread of organized fraud rings costing carriers and consumers in excess of $120 billion in reserve exposures, state mandated anti-fraud programs are no longer the exception. They are now a necessity. While traditional forms of Special Investigative Unit's (SUI) dominate the claims community, the sophistication, technical and legal prowess of fraud rings, organized crime, and plaintiff provider scams necessitates that claim units consider a dual track approach to protecting reserve exposures with prioritization placed at the front end of a claim rather than in the midst of costly reserve exposures.

According to the FBI's Economics Crime Unit there are currently 5,000 insurance companies with $1.8 trillion in assets. The industry's two largest components are identified as property / casualty and life / health. According to the FBI, insurance is a "critical industry" and falls within the defining parameters of "Tier One" of the FBI's "Strategic Plan."

According to a preliminary study conducted by the Washington DC based, Coalition Against Insurance Fraud (CAIF), quoted in the March 2000 edition of Claims Magazine, there are now 36 states with 44 of their own fraud bureaus employing 1,158 personnel, 64% of which represent investigators. In addition, 44 states now define insurance fraud as a felony.

In 1997 the Insurance Research Council, which conducted a survey of 150 insurers, concluded that "direct spending on fraud detection and prevention grew significantly from approximately $200 million in 1992 to at least $650 million in 1996." The survey found that in 1992 insurers that operated SIU's serviced 66% of the property casualty market. By 1996 this figure had grown to 76%, and increases were planned through 1999.

Conversely, CAIF has estimated annual losses of $27.6 billion from insurance fraud, not inclusive of health-related losses. Basic estimated losses by lines: auto: $12.3 billion, homeowners: $1.8 billion, business/commercial: $12 billion, and life/disability: $1.5 billion.

A study conducted by Conning & Company, a broker-dealer subsidiary of Conning Corporation registered pursuant to the Securities Exchange Act of 1934, revealed that $95 billion was lost in the arena of health care fraud reflecting a staggering 80% of the total of all losses against all lines. The Conning study went on to cite that WC losses stemming from fraud were 25% of established reserves. Aggregate losses reflected an estimated $120 billion in a 1995 study across all lines that included all private and federal programs.

Contemporary Trends
Joseph Cohen, chairman of the Anti-Fraud Task Force for the National Association of Insurance Commissioners, told the National Underwriter in an interview, "I'm starting to get the feeling that some companies are starting to pull back on anti-fraud efforts. When you see companies outsourcing [fraud investigators] it indicates they don't have the employees within the company." Both manpower and resource reductions within the context of recent budgetary realignments have no doubt impacted SIU operations.

Given the celebrity status given to state-organized "insurance fraud bureaus' and the publicity surrounding their successes, other state operations, with the exceptions of New Jersey, New York, and California, are also hampered by limited manpower and resources. For example, New Jersey's 1998 operating budget rose 233% in 1998 to $28.5 million dollars and reflected 30% of the aggregate operating expenses for the remaining 43 state fraud bureaus.

Another trend making its impact felt is the increase of bad faith lawsuits stemming from improperly trained investigators, claims misidentified as being fraudulent, and the improper or over-investigated claim and its resulting outcomes.

Foundational Issues for Anti-Fraud Programs
Given the potential for losses stemming from fraudulent claim filings, the necessity for implementing a comprehensive and effective anti-fraud program is paramount. As part of an anti-fraud development program priorities should be established in developing a two-tiered operational approach. Preventive interdiction programs designed as first line detection systems in preventing initial reserve exposures, as well as SIU fraud programs, should comprise the two-tiered anti-fraud program. Preventive interdiction programs also serve to enhance pre-SIU fraud referral capabilities. Implementing such programs requires identifying available resources, manpower, effective training, implementation, support, and establishing a mission statement that can be fulfilled.

Additionally, whether operating in a multi-jurisdictional or in a single jurisdictional environment, it will require knowing the standards of those states requiring a compulsory anti-fraud program and a reporting compliance.

In discussing the viability of setting up or enhancing your anti-fraud program, it is imperative to understand the defining parameters of fraud and evidentiary standards required to initiate legal action against parties involved in fraudulent activities. By understanding both these concepts you can better establish where resources should be allocated and where improvement measures within an existing anti-fraud program need to be focused.

Because there are 50 states that comprise the United States, there exists a concurrent set of case law and statutes for each of these states that define fraud and establish evidentiary standards for proving fraud. Within this context, we will discuss the fundamentals governing the definition, detection, and consideration for legal action for those claims demonstrated as being fraudulent.

In discussing the formation or enhancement of anti-fraud programs there is a concomitant proprietary reality that must be calculated into the equation when establishing claim resolution and settlement protocols. This would be the bottom line. Insurance companies, self-insured entities, and risk management organizations are not motivated to establish anti-fraud programs exclusively for altruistic reasons. Saving money, increasing profit margins, containing operational costs, and stabilizing premiums paid by the consumer factor into the anti-fraud development program.

As with any successful organization, change, evolution, and mission statements must be subject to amendment in response to the changing state of information technologies, statutes, regulations, operational capabilities and enhancements, and ethical compliance issues. In response to current advancements and legislative changes the claims community must now give consideration to developing anti-fraud programs within the scope of a two-tiered approach. Preventive interdiction and SIU fraud operations, as part of a two-tiered anti-fraud program is key to protecting established reserves and preventing initial exposures as a result of a passive claim resolution and settlement philosophy.

Defining Fraud In A Legal Context
The subject of insurance fraud is multi-faceted. Defining the term "fraud" brings with it the nuances and subjectivity ascribed to the legal issue of pornography. The most commonly attributed statement associated with the discussion of insurance fraud has been, "I may not be able to define it, but I know it when I see it." From this obviously subjective standard comes an issue related to how anti-fraud programs are established, their philosophies,

mission statements, zealotry, training programs, and bad faith lawsuits.

So what is fraud? In trying to define "fraud," a more suitable and legally recognizable approach is to discuss what basic elements are considered to constitute fraud. Typically, there are six recognized elements of fraud:

1. A representation
2. It's material to the claim
3. It's made falsely with knowledge of its falsity or recklessness as to whether it is true or false
4. It is done with the intent to mislead someone else to rely on it
5. There is a justifiable reliance on the misrepresentation
6. A resulting injury is caused by the reliance
7. Evidentiary Requirements

The question of actual fraud is one of fact and has to be determined by a jury. However, whether the evidence meets the required standard of clear and convincing proof to justify its submission to the jury is a question of law for the court. Fraud is generally not to be presumed and must be established by direct proof or by facts clearly proved and sufficient to warrant a presumption of its existence.

- Fraud must be established by evidence that is "clear and convincing"
- Fraud may be shown by circumstantial evidence, as well as by direct evidence, or by a combination of the two

Elements of Misrepresentation

In certain states negligent misrepresentation is a recognized claim and can be alleged in lieu of lawsuits for fraud. Negligent misrepresentation requires proof of:

- Misrepresentation of a material fact
- Misrepresentation made under circumstances in which the actor should have known of its falsity
- Misrepresentation is made with the intent to induce another to act on it
- Injury to a party who justifiably relies on the misrepresentation

Fraud, Misrepresentation and Non-Disclosure

Under NAIC provisions, regulated institutions involved in the business of insurance maintain regulatory and statutory rights of investigation. Earlier we addressed the nuances and subjectivity surrounding the issue of fraud. In developing anti-fraud programs organizers must be keenly aware of other impacting claim issues that determine whether anti-fraud measures are warranted, and whether significant fraud indicators are present and discriminate between definitive fraud indicators and issues related to misrepresentation and non-disclosure, which may not be directly attributed to legally definable fraud.

Keeping in mind previously mentioned elements of fraud and the evidentiary threshold required to pursue legal action, anti-fraud programs should have in place mechanisms which will allow investigative units to gauge issues related to misrepresentation and non-disclosure as measured in terms of intent, willfulness, and knowledge. In citing an example of how issues related to non-disclosure do not automatically trigger certain fraud indicators, typically 25-32% of all applications for PIP/UM benefits filed through a carrier by first and third parties end up transferring the financial liability to other carriers disclosed, only after initiating an eligibility investigation vis-à-vis independent investigative services. Using a baseline nuisance value of $8,000 per claim, the protected aggregate reserves per 1,000 claims are estimated at close to $1 million. The issue of non-disclosure in the majority of claims filed in this manner are a reflection of the consumer's ignorance of the laws governing the settlement claim practices within the state affected and not necessarily from willful or intentional fraud.

Another example of unnecessary reserve exposures due to non-disclosure has to do with Worker Compensation claims filed in states where second injury funding provisions exist. This is a multi-faceted issue that has gone relatively unaddressed. Claim units operating in jurisdictions where second injury funding provisions exists typically fall short of the due diligence threshold necessary for establishing baseline requirements for determining whether a particular claim is subject to the second injury funding provisions.

Medical indexing searches of claimant histories are inadequate for establishing prior and similar pre-existing injuries. Exercising other available and comprehensive investigative options should be a requisite in the claims resolution process pursuant to injury funding provision requirements.

The Bottom Line

Without diminishing the dedication, commitment, hard work, and professionalism demonstrated by SIU units throughout the U.S., every corporation is driven by a single imperative—profit margins. The insurance industry is no exception. In addressing the issue of establishing anti-fraud programs the profit factor is central to the operation's mission statement and claims resolution philosophy.

Understanding that the bottom line is the bottom line to every aspect of the insurance industry as a business, organizations operating within the parameters of a passive claim resolution philosophy are marginalizing the ability to protect reserves, detect fraud, and enhance the financial viability of the institution they represent.

Once again, citing the statistical information regarding PIP/UM eligibility determinations, as well as those referencing second injury funding provisions, claim operations which choose to settle claims arbitrarily without a requisite minimal inquiry are potentially exposing reserves which otherwise might have been defended and are causing unwarranted losses.

Current advances in a number of areas—information technologies, the expanded jurisdictional scope of investigative inquiry, enhanced and comprehensive methods of identifying fraud, misrepresentation, and non-disclosure, the increased cost of settling claims at nuisance value, the integration of proprietary fraud examiners skilled at detecting fraud, providing reserve protection at the initiation of the claim, and, stemming the intervention of costly litigation—it is imperative that insurance organizations restructure their anti-fraud programs to incorporate a two-tiered fraud prevention and detection system.

Anti-Fraud Plan Guidelines

While anti-fraud programs and reporting requirements vary from state to state, there are basic guidelines shared by most anti-fraud plans. Most structured anti-fraud plans are ultimately implemented through special investigation units.

While not typically a requirement of most compulsory anti-fraud plans, most insurance companies develop a mission statement which goes to the heart of the anti-fraud effort that the companies implement. Most anti-fraud plans may also include the primary contacts or directors of anti-fraud operations within the organization.

For internal purposes an organizational chart and professional dossier on all members of the SIU unit should be maintained in the event that professional bona fides need to be provided on demand by state fraud bureau officials or designated fraud prosecutors.

While the scope of most discussion forums almost exclusively address the issue of external fraud or what most of us in the anti-fraud community refer to "claimant" fraud, a comprehensive anti-fraud plan must address both internal and external fraud prevention measures.

Six Primary Guidelines

1. Fraud Prevention & Detection Procedures: These are procedures to prevent insurance fraud, including internal fraud involving employees or company representatives, and external claim fraud resulting from misrepresentation on applications for insurance coverage.
2. Fraud Review & Investigations: This part of the plan encompasses the review of claims in order to detect evidence of possible insurance fraud and procedures for investigating claims where fraud is suspected.
3. Referral of Fraudulent Activity to Law Enforcement: The insurance organization must have a systematic method of reporting fraud to the appropriate law enforcement agencies, fraud bureaus, and the designated prosecutor's office.
4. Remedies Against Fraud: This aspect of the fraud plan requires carriers to develop plans and procedures dealing with restitution or other damages either through house counsel, independent counsel, or under criminal proceedings.
5. Fraud Detection Training Plan: This portion of the fraud plan includes all aspects of anti-fraud training, including both internal and external programs.
6. Fraud Warnings: Pursuant to the NAIC's and/or CAIF's standardized statements of warning, insurance documents should include "substantially similar" warnings to the effect that: "It is a crime to knowingly provide false, incomplete, or misleading information to an insurance company for the purpose of defrauding the company. Penalties include imprisonment, fines, and denial of insurance benefit."

Anti-Fraud Plan Outline

I. Establishing Tasks, Conditions, and Standards
   a. Requirements for two-tiered anti-fraud program
   b. Defining fraud
   c. Establishing the six guidelines for an anti-fraud program

II. Introduction
   a. Global perspective of insurance fraud
   b. Statistical data concerning industry losses
   c. Contemporary trends
   d. Establishment of fraud bureaus

III. Fraud
   a. Fraud as a legal definition
   b. Elements of fraud
   c. U.S. Federal Court citings
IV. Evidentiary Requirements
   a. Fraud as a question of fact
   b. Standards of proof
V. Misrepresentation
   a. As a legal defense
   b. Standards of proof
VI. Fraud, Misrepresentation, and Non-Disclosure
   a. Understanding the differences
   b. Incorporating into an anti-fraud plan
VII. Foundational Issues for an Anti-Fraud Plan
   a. Two-Tiered Operational Program
      1. Preventive interdiction program
      2. SIU operations
   b. State mandated anti-fraud plans
   c. Altruism vs. profit lines
VIII. The Bottom Line
   a. Insurance as a business
   b. Mission statements
   c. Reserve protection
IX. Anti-Fraud Plan Guidelines
   a. Introduction
   b. External vs. internal fraud plans
   c. Six basic guidelines
X. Red Flag Indicators of Fraud
   a. Worker's compensation
   b. Health insurance
   c. Auto
XI. Regulatory, Statutory, and Ethical Compliance Issues
   a. GLBA
   b. NAIC Model Provisions
   c. FCRA
   d. DPPA
   e. Statutory provisions
   f. Vendor certification
   g. Procuring personal and privileged information
XII. Investigative Applications
   a. Information technologies
   b. Proprietary investigative options

Jeffrey A. Russell is President and Founder of National Claims Information Systems, Inc. He is a member of numerous professional groups, including the Association of Certified Fraud Examiners and the National Society of Professional Insurance Investigators.

How Some Terrorist Groups Raise Dirty Money in the U.S.

By Don deKieffer and Todd Sheffer

The terrorist attacks have not yet been fully investigated, so it is too early to speculate as to how these grisly operations were financed.

What is clear, however, is that terrorists and their sympathizers have long used commercial fraud in the United States to raise funds for their activities.

Although no one is sure of the extent of these schemes, a few examples illustrate their *modus operandi*:

- The 1993 World Trade Center bombing was financed in large part by an elaborate coupon fraud and counterfeiting conspiracy. Mahmud Abouhalima, Ibrahim Abu-Musa, and Radwan Ayoub established a network of stores in New York, New Jersey, and Pennsylvania, which were later identified as having associations with persons identified as the WTC bombers.
- Adnand Bahour, the nephew of George Habash (leader of the Palestinian Liberation Front) set up a fraudulent coupon distribution network through his grocery stores in the Hollywood, Florida area. Bahour was a kingpin in the national terror network creating money laundering and financing for the PLO. During the investigation and subsequent raid on the meeting hall of this network in Hollywood, more than 72 individuals from throughout the United States gathered to further their fraudulent coupon distribution network.
- The Secret Service has reported that a massive credit card scam was being perpetrated by a group of Middle Easterners with affiliations to known terrorist groups. This scheme was a "Regulation Z" fraud or "Booster Check/Bust-out" scheme. The investigation revealed that the Middle Easterners, organized into cells located throughout the U.S., had applied for and received numerous credit cards. These cardholders systematically "boosted" the credit limits of the cards to the maximum amount available. Once they had established their portfolio of unsecured credit card debt, they submitted worthless checks as payment for these accounts in advance of purchases being made. In most

cases, the checks were in amounts exceeding the cardholder's credit limits. Before the checks were returned as worthless, the card holders managed to purchase merchandise and obtain cash advances up to and sometimes in excess of the limits on the accounts. Losses to banks and merchants from these frauds were more than $4.5 million. Many of the fraudsters subsequently fled the country.
- Several Islamic charities, which solicited funds and products from U.S. companies, are little more than fronts for terrorist organizations. Examples include the Wafa Humanitarian Organization and the Al Rashid Trust.
- Numerous companies in the Middle East and elsewhere have been identified as agents for terrorists and/or states that harbor terrorists, such as Iraq and Libya. Several U.S. firms are known to have had transactions with them. For example: *Iraqi Sanctions Regulations.* Tigris Trading, Inc., London, England; Dominion International, England; Bay Industries, Inc., California, U.S. *Libyan Sanctions Regulations.* Corinthia Group of Companies, Floriana, Malta; Oil Energy France, France; Quality Shoes Company, San Gwann, Malta.
- The FBI smashed a massive cigarette smuggling ring in Detroit and arrested 19 people. The proceeds from this operation were being shipped to Hamas and other terrorist groups. The smugglers purchased large quantities of cigarettes in low-tax North Carolina and trucked them to Detroit where they sold in Arab-owned convenience stores.
- A group of 14 Lebanese and Syrians were convicted on dozens of charges for operating a multi-state ring that repackaged stolen and counterfeit infant formula. Much of the money acquired by this scam simply vanished, although there are some indications it was sent to the Middle East. The leader of the ring, Ibrahim Elsayed Hanafy, faces a maximum prison term of 623 years.
- Investigators have linked relatives and/or associates of longtime business operators in California, New York, Michigan, Illinois, Florida, and Washington to bust-outs, phony checks, cigarette tax schemes, diversion fraud, credit card fraud, counterfeit consumer products (including baby formula), insurance fraud, and food stamp fraud. Investigations uncovered large amounts of cash wired from the United States to the Middle East, and many of these frauds were conducted with paramilitary-like precision.
- Billions of dollars in orders for U.S. goods have been shipped to trading companies in the United Arab Emirates (U.A.E.) especially Dubai recently. More than 70% of all

goods shipped to the U.A.E. are subsequently exported, often to sanctioned countries, such as Iran.

What Can Companies Do About This?

1. The most obvious, but frequently ignored method of de-funding terrorists is to *know your customer*. The government regularly publishes lists of "denied parties" and updates its web sites. Export Administration Bureau can be accessed at www.bxa.doc.gov and Foreign Assets Control can be accessed at www.treas.gov/ofac. These sites are somewhat difficult to use, and you must use both to get a comprehensive list. A better on-line resource is Trade Compass at www.tradecompass.com, which combines many publicly available lists.

   The Washington D.C.-based law firm of deKieffer and Horgan maintains what is believed to be the most comprehensive database extant of known diverters, counterfeiters, smugglers, and their accomplices. It also keeps files on companies and individuals identified as terrorists (or their surrogates) by the U.S. government. The database, which is called EDDI, took over ten years to build and has over 160,000 records. EDDI's on-line service is an unlimited resource for businesses to screen potential and current customers. For more information on EDDI call 301-829-9384 or email ddekieffer@dhlaw.com.

2. Be able to track your products. Detecting counterfeit and diverted goods is essential to knowing if you have a problem. There are several reputable companies, which can help with packaging and marking technologies.

3. After consulting your attorneys, contact law enforcement authorities such as the F.B.I if you suspect that you have had dealings with a suspicious individual or company.

U.S. companies must do their part to de-fund the terrorists. This is not business as usual. Being a victim of terrorists and funding them at the same time is not only stupid, but it can be deadly.

Donald deKieffer has practiced international trade regulation law for more than 25 years. After graduating from the Georgetown University Law Center, he worked for the Senate Republican Party Committee as a foreign trade specialist. In early 1980s Mr. deKieffer served the Executive Office of the President as General Counsel to the United States Trade Representative in Washington, D.C. He is a frequent speaker on international trade regulatory issues and is the author of five books and more than two hundred articles.

Todd Sheffer is the National Special Investigation Unit (SIU) Manager for Magellan Health Services, the world's largest behavioral health care company. He is also an active member of the Maryland Chapter of the Association of Certified Fraud Examiners and the International Association of SIU.

Chapter Five

Collections

Collecting on Civil Judgments
By Christina Smiley

Did anyone ever owe you money? Of course they have. Many of us at some point have encountered a situation where money is owed and never repaid. After exhausting all of the traditional methods of bad debt recovery at our disposal, it then becomes a matter of using our legal system to further pursue the collection of the debt. For a very lucky few, after winning the judgment the debtor will pay and that will be the end of it. Unfortunately, for most of us a civil judgment ends up being nothing more than a useless piece of paper. And you're not alone; nearly 80% of all civil judgments remain unsatisfied. Unless you understand how to use post-judgment remedies to enforce the decision and also know where the assets are, after the final bang of the judge's gavel it may all have been for nothing. Fortunately, if you've won a judgment, there are many powerful legal tools at your disposal to force an unwilling debtor to pay. For your amusement here is a typical small claims case scenario.

The Dispute
Imagine yourself as John Doe. Now you're a nice enough guy and one of your less fortunate friends, Dan Willnotpay, approaches you for a loan so he can buy a used car. Let's say it's $2,000 that he wants to borrow. Being of a generous nature, you loan him the money with the understanding that he'll pay you back in two months.

All right, so two months have gone by, and your friend has been mysteriously absent. You've called his house and left several increasingly anxious messages—which he doesn't return. Just to make the story more interesting, imagine you've heard from a secondhand source that Dan has received a hefty sum of cash from a recently departed relative, so you *know* he has the money to repay your loan.

Now six months have gone by and Dan seems to be doing well enough, but he still won't pay you back. You do what the average law-abiding citizen does and take the son of a gun to Small Claims Court. And besides, you can really use that money to put braces on your kid, John Jr.

The Court Clerk and the Judicial Court System
After paying your filing fee you are met with an onslaught of strange looking forms that must be completed before you can proceed. After you take these forms home and puzzle them out, you

return to the court and hand them over to the court clerk who promptly tells you they're not filled out correctly and hands them right back. It's a given that you're confused by this point so you ask the clerk to explain where the problem lies.

Now, getting information out of a court clerk can be like pulling teeth if you don't know the right questions to ask. So likely if you get any information at all, you may be more turned around than when you asked in the first place. But somehow you manage to get it done correctly and you hand in your paperwork and schedule your court date.

Your day in court finally arrives and—wonder of wonders—Dan doesn't even show up. But that's not necessarily a bad thing because the judge has awarded you the judgment by default. You breathe a big sigh of relief and go home waving in triumph your piece of paper that says you were awarded the judgment. You tell your wife to go ahead and make that orthodontist appointment for John Jr.

A month goes by. No word from Dan. No word from the court either. You're thinking there must have been some hiccup in the legal process and that's why you haven't received a big, fat check in the mail yet. Naturally you call the court clerk—that lovely fountain of information—to ask why you haven't received your check yet.

Now comes the big surprise.

The real whopper.

It's totally up to you to make Dan pay. The court can't help you. So sorry. Have a nice day.

In most states you can garnish Dan's wages, seize his bank account, or even put a lien on the title of that used car, but do you know how? What if he's working under the table? How can you get a piece of *that* pie?

Back to square one, which is nowhere by the way. It seems like pigs are going to fly before Dan pays up on his debt.

For most judgment holders this is the end of our tale. Without any real knowledge about how to enforce a judgment it usually ends up being nothing more than a useless piece of paper taking up space in a file somewhere until it expires. Fortunately this need not be the end of our story for those who are willing to wade through the murky waters of judicial judgment enforcement. For those who endeavor to pursue, the real work now begins.

The Skip Trace

The chase is on! Since John didn't really have very much information about Dan's assets, it is now up to him to find them in order to seize them to satisfy the judgment. There are many means of asset discovery available to judgment holders. One of the strongest tools available would be a full consumer credit report. A judgment creditor has what is referred to as "permissible purpose" to obtain this information on the judgment debtor. In most circumstances a consumer credit report will provide an enormous quantity of information about the debtor's current and previous addresses, assets, and financial situation. In addition to this it will also help to provide useful leads to *other* people and companies who may have information about your debtor.

Another useful procedure is a debtor's examination. You may compel a debtor to appear with you in court to answer questions about assets and other financial matters under oath. Aside from the most obvious benefits of a debtor examination, also consider how terrifying it is to be ordered to appear in court to answer your questions! Failing to appear can result in a bench warrant issued for the judgment debtor's arrest. Above all else, the debtor examination can be considered a terrific legal intimidation tool. In addition, a "Turn Over Order" may be used at the examination to seize any assets, including cash, jewelry or other items on the debtor's person. A "Turn Over Order" is exactly what it sounds like—an order to "turn over" specifically named assets.

It must be said that some assets are exempt from seizure for the enforcement of a judgment. These assets that are exempt from seizure and the enforcement procedures themselves tend to vary slightly from state to state. So before going further, I'd recommend that you find out what assets are not exempt as well as become familiar with the enforcement procedures in your state. You'll find that there are many resources on the Internet to research your own state's statutes and code.

Typically, the following assets are among those which are *not* exempt and should be considered: wages (25%); automobiles; recreational vehicles; bank deposit and savings accounts; income, property or interest due from a third party; business income; and real or personal property. Some assets, like real and personal property, will carry partial exemptions. For example, in California you may seize an automobile for sale to satisfy a judgment, but if it's a primary vehicle the first $1,900.00 of equity in the vehicle is exempt. Some of the "biggies" that you *can't* go after would be family support (child or spousal), public assistance (welfare), Social

Security, disability, worker's compensation, and retirement benefits. This is by no means a complete list of exemptions, so be sure to become familiar with all of them.

The Enforcement Process

Back to the story: After locating Dan's bank account and employer, John got down to business. Of course, he started by going after the easiest, biggest asset available—the bank account. He filed a form called a Writ [commonly referred to as a writ of *execution* or a writ of *fifa*] with the court and instructed the Sheriff, who acted as his *levying officer*, to go over to the bank and seize the money in the account. Note that you always direct the Sheriff to seize assets *in the county where the assets are located.* For example if the judgment was awarded in El Dorado County, but the bank account is located in Sacramento County, you would first obtain a writ from the court in El Dorado County and then direct the Sheriff in Sacramento County to seize the assets. This being the legal system, after all, it took about 30 days for the Sheriff to complete the seizure and issue John a check.

Any allowable fees that John spends are reimbursable as costs after judgment, so John is able to add the amount of these filing and levying fees with a Memorandum of Costs. Note that in some states this is accomplished with a declaration or an affidavit. Dan will ultimately have to pay for them.

Now you reap what you sow after all, and Dan had been a little sleazy, so Dan only had about $1200 of his settlement left in his bank account. That left a balance due on the judgment of $842 ($800 of the original judgment + $42 in reimbursable fees). No problem for John. He found out where Dan works, remember? John once more obtained a writ, then sent the Sheriff to do his work for him and serve a garnishment order on Dan's employer. Now he'll get paid every time Dan gets paid until the judgment is completely paid off.

For those of you with financial savvy, I haven't included post-judgment interest in this story just to keep things uncomplicated, but in reality interest accrues on Dan's debt at a simple interest rate from the date the judgment was awarded, and that increases the amount of the judgment.

In any case, the ending to this story is a happy one. John got all of the money that Dan owed him, plus costs and interest, and John Jr. got those braces he really needed. We don't really feel very sorry for Dan.

Now I realize that this by no means will be a resolution for every situation. Fortunately for judgment holders there are many other remedies available to satisfy judgments. It also bears mentioning that a few states don't permit certain procedures. For example, wage garnishments are not permitted in North Carolina, South Carolina, Texas, and Pennsylvania, to name a few states, except in limited circumstances, such as the enforcement of child or spousal support. Also, it isn't permitted to garnish the wages of anyone claiming a "Head of Household" status in Florida. Bank account levies are not permitted in Delaware. The only way to determine what is and what is not allowed in your state is by conducting your research on a state-by-state basis.

It goes without saying that in most states you can seize property in order to satisfy your debt. However, seizing property can be an expensive and time-consuming endeavor. Nevertheless, sometimes this will be your only option and may be useful even if your goal is not to actually sell the property, but to *make your debtor pay*. If your debtor possesses an enormously large ego, you might want to seize his brand new Corvette or his classic muscle car. If your target runs a successful data processing business, you might want to seize the main computer server, effectively bringing the debtor's business to a screeching halt. Starting to get the drift?

Before proceeding you'll want to determine the amount of equity actually available in the property you're planning to nab, the cost involved, and whether or not it's worth your while to go through with it. Your due diligence when making this determination should include contacting the County Recorder in the county where the property is located to find out if there are any senior liens ahead of you that will allow for any partial exemptions of the equity in the intended asset as well as deducting any balance due on a loan or mortgage. When property is seized for sale, it must be stored in a proper facility until the sale takes place, and this requires a hefty deposit. After the property is seized, the debtor will be notified that a sale will take place unless he or she coughs up the money to pay you. If a sale does proceed you'll usually get about ten cents on the dollar of the actual property value.

A long-term alternative to ensure payment of your judgment would be to secure a lien against any property the debtor might own. Liens are affected by having the court where the judgment was awarded issue an Abstract of Judgment, then securing the lien by recording the abstract with the County Recorder in the county where the property is located. Liens are enforced at the time the owner of the property either attempts to sell or refinance the property. They are honored in the order that they are secured, so if a lien was placed in the same county ahead of yours, any equity in the property goes to the lien ahead of you, and whatever is left goes toward satisfying your judgment.

Should your debtor be fortunate enough to own a successful business, you can use either a Till Tap or a Keeper to aid in the enforcement process. Both of these procedures are useful when the debtor's business is some form of retail where customers are shopping on the premises. In the case of a Till Tap a levying officer (sheriff) at your instruction will appear at the business and collect any and all money found on the premises. A Keeper can be used to collect the money on the premises and then stay behind the cash register for eight or more hours. If your debtor has a high profile business, this is sure to send him or her running for the checkbook! A word of caution: Though it's not commonly known by the average deadbeat, the debtor can close the business for the day to avoid either a till tap or a keeper, thus forfeiting your deposit in the process.

Let's assume your debtor has income, but it's not traditional income. Perhaps she's getting her cash under the table by babysitting. Maybe he has income from a rental property or has the type of business that makes it difficult to use a Till Tap or Keeper to intercept income. Try to think outside of the box on this one, but keep it simple. Find the money—the actual source—and intercept it. She's babysitting? For Whom? Mrs. Jones? Then Mrs. Jones is your target He's doing landscaping? Who for? Joe's Motel. Then Joe's Motel is your target. If there's a rental, target the tenant and intercept the rent. The interception of money due to a debtor by a third party is referred to as a Third Party Levy. This type of levy orders the person/company/whatever, which has money that belongs to your debtor to hand it over to you instead. This works for just about any stream of income as well as property in a third party's possession.

These are some of your more basic methods of recovery. If you should ever find yourself in the unhappy position of being owed a debt and faced with the reality of pursuing payment through legal channels, I have a small measure of advice for you. Arm yourself with as much knowledge about the incumbent debtor as you can *before* you ever find yourself in this position. If you're going to lend money, write up an agreement and get financial information and a signature. If you have rental property make a prospective tenant fill out a thorough rental application and *verify* the information before renting. Make copies of any checks you receive. Keep copies of canceled checks that you have written since these will reveal the bank account number of whoever cashes them on the back. Protect yourself ahead of time, and you'll save yourself a lot of grief down the road. In the end you, or at least someone you know, will end up with an unsatisfied judgment. If you already know where the assets are you'll be ten steps ahead of the game.

Christina Smiley is the founder and owner of Sierra Judgment Recovery, located in South Lake Tahoe, California. Sierra Judgment Recovery has been actively enforcing judgments in California and providing professional judgment recovery training since 1994. Ms. Smiley was a founding Board Member for the California Association of Judgment Professionals and is also their Director of Continuing Education. She is also a member of several professional judgment recovery associations. Her primary interests are education of new judgment recovery specialists and the further development of the judgment recovery industry.

Post Office Box Process Service
By T. A. Brown

If you are trying to have legal documents served on someone, and all you have is a post office box for an address, you will need to be able to obtain the physical address records from the post office or mailbox business. Privacy laws prevent giving this information out to just anyone, but if you have legal documents that have been filed with the court for process service, the post office box provider must supply you with the address they have on file. Some post office or mailbox businesses have personnel who are reluctant to give this information out and they require proof that the law demands it of them. Consequently, to save time and frustration, it is better to go prepared with a letter that states the laws in your state that will make it clear to the mail box provider that you have a legal right to the information requested.

There is a self-generating letter on the Internet that you can print out and provide to the post office box carrier in order to obtain box holder information, using your own information and the laws of your state. If you are using IE5 or better, just go to: www.isitserved.com/dforce/anypostal.php3

The name and last known address are required for change of address information. The name, if known, and the post office box address are required for boxholder information.

The information is provided in accordance with 39 CFR 265.6(d) (6)(ii). There is no fee for providing boxholder information.

Unmasking the Mystery of Hidden Assets

By Kelly E. Riddle

The thought of conducting a hidden asset investigation can send shivers up the spine of many a good investigator. The word "hidden" almost conjures up a thought process that the assets can't be found because they were intentionally hidden. Sometimes this is true, while other times the assets are simply not in a convenient place for us to find them. There are a variety of reasons for conducting hidden asset investigations, and these types of investigations are great for cross-selling. For instance, if you were hired to check the activities of your client's spouse and you conclude that the spouse is having an affair, you should suggest to your client that they conduct a hidden asset investigation. This benefits your client by giving him or her a total picture of his assets and benefits the investigator by providing more work. I have been involved in numerous cases where the hidden asset investigation turned up more than one-million dollars in assets that the spouse did not know about prior to the investigation.

There are a variety of other reasons a client may have you conduct a hidden asset investigation. Most of the time this is done in an attempt to satisfy a delinquent debt or judgment. Insurance companies conduct hidden asset investigations called "subrogation" whenever they pay for the damages their insured received because the other party is uninsured. The insurance company then tries to locate assets to reclaim the money spent on their insured. Businesses may conduct hidden asset investigations before entering into business dealings with another company or a person to make sure they are stable. Regardless of the reasons, hidden asset investigations can be a good source of income for an investigator.

During the initial consultation many clients will often request that the person's credit report be obtained. Obviously this can't be done legally unless you have the person's written permission or there are certain other criteria. At this point, you know that the client is trying to determine what the person's financial picture is, and I typically advise the client that we can provide a "financial profile." Although we are unable to access their credit report, other records can give you information about the person's financial status. For instance, if you check the district civil and county civil records and find that the person has several lawsuits against him for failing to pay debts or failing to pay taxes, you can assume that this is a characteristic of that person. By building on this, you can provide the client with an overall profile of the level of the person's financial responsibility.

Conducting a hidden asset investigation is simply following the basic rules of investigations. You have to start with the known information and work towards the unknown. The investigator should also understand civil laws in his or her state to ascertain whether or not time and money should be spent investigating other subjects who may be hiding assets for the subject. For instance, is your state considered a community property state? If so, the person probably would not hide assets in the name of their spouse since half of the asset is already theirs by law (assuming they are trying to hide assets during a divorce). Most states have guidelines regarding transferring property into another person's name prior to a bankruptcy, judgment or other related civil action. Normally, if the assets were transferred within six months to a year of the civil action, they are considered part of the civil action. Part of the investigator's job is keeping the client focused and reasonable. It is not uncommon for a client to provide a list of 20-30 people they think could be hiding assets.

Obviously the cost of checking each of these people would be enormous and probably futile. The investigator must therefore encourage the client to provide only those names that are prime candidates for hiding assets and to think of people in the subject's past whom they trust. A lot of who the person uses to hide assets depends on the value of the asset and the type of asset. If the asset is a large boat or airplane, they may have the vessel taken out of service for maintenance and work out an agreement with the mechanic for storage. The person may have a small amount of cash, and this is typically given to a confidant to hide for them. Large amounts of cash are placed in off-shore accounts, structured investments, and related hiding places.

The investigator should start with known information, such as the person's current address, name, and social security number. All addresses should be checked through the appraisal district to determine the actual owner of the property. If the subject is found to own the property, the investigator should check the records to determine if there is a mortgage holder or lien holder listed on the property. Even if there is, a phone call should be made to the institution to make sure that the note hasn't been paid off and the property is free and clear. Remember that the first three numbers of a social security number tells you which state the number originated in and therefore guides you to another geographic area to investigate.

I had a case where a bank loaned a subject from Louisiana $750,000 to build a strip-center-type mall. The subject never built the mall and hid the money. I tracked the subject's asset trail all over Louisiana and Texas and was unable to locate the money. While combing over the file for that piece of evidence that would open the door, I realized I had skipped the obvious. I checked his social security number and found that it originated in North Carolina. After conducting some quick record checks, I discovered that he still had a mother and brother living in North Carolina and soon located the bank where he had hid more than $500,000.

The person's movements will provide insight into where assets may be hidden. By checking their social security number, driver's license history, determining where they travel, and related information, you may soon discover the location they will most likely use to hide assets. Another useful tool is the good old tool of "dumpster diving." In a recent case we were hired to check the financial status of a subject who owned a real estate company and a custom home building company. One night we gleaned through the trash in his dumpster and found reports detailing all of the houses constructed over the past three years—information about current construction projects, total expense, total profit, number of properties listed by the real estate company, total revenue generated over the past three years, a list of employees, telephone records, and much more. Of course, the investigator should keep in mind trespassing laws. In this case the dumpster was in an office complex readily accessible.

Public records should be researched for any clues related to assets and should include the following:

1. District Civil records: Check for cases involving divorce, debts, judgments, damages, auto accidents, business disputes, and related cases. Even if some of these cases are old, pull the file and research the enclosed information. In many of these cases the subject has to prove his assets or damages and provides financial-related information. In one case when I reviewed an old divorce case, I found that the subject raised Arabian horses. Armed with this knowledge and the Internet, I was able to find his web site and determine where his current ranch and stables were located.
2. County Assumed Name records: A check of these records will provide information about any alias names or "doing business as (dba)." If you find a listing for a business, review the actual record to determine what address the subject used and if he listed any partners.
3. County UCC (Financial Statements): If a person obtains a loan and posts any type of collateral, this information is

often filed in the county financial the asset if free and clear and still owned by the subject.
4. County Deed Records: These records deal with the purchase and sale of property. However, in most jurisdictions much more is included in these records. Other information often included in deed records include gifts, garnishments, power of attorneys, liens, judgments, mineral rights, etc. Even though the record may indicate that the person sold the property, the investigator should take note of the purchaser of the property since this may be a friend who is being used to hide the property. The bank involved should also be noted since this may tell you where the person does most of his banking.
5. County Brand Indexes: Most counties require livestock brands to be registered. One case that I worked was successful because I took time to check this index. I found that the subject had a registered brand, and I called a local feed store in his area and was able to determine where the subject had his cattle. I drove to the area and found oil wells pumping on the land. I took photographs of the cattle and the oil wells as well as of the related equipment on the land.
6. County Tax Assessor: Check the tax assessor's office to determine any property the subject may be paying taxes on, including automobiles. This doesn't necessarily mean he owns the property simply because he is paying the taxes, but logic would tell you that he has some kind of financial tie to the asset.
7. County Appraisal District: I use this to cross-check information from deed records and tax assessor's records. If property is found, the appraisal district assesses the property's value for taxing. Be sure to check the record to determine if the mailing address for the subject is different than the property address.
8. Local Police Department: Run a "name survey" through the local police department for at least the past year. Ask for a list of every call associated with the person's name. You may discover that the police were called to a dispute at a rental house or other location that the subject may have financial ties with. If a record is found, pull the actual document to determine what assets were posted. Next, follow up to see if the note was paid off.
9. State Comptroller's Office: The records will provide information about any businesses associated with the subject, mineral and franchise assets, whether the subject is or was a state employee, and related information.
10. Secretary of State: The record will provide information related to any businesses associated with the subject.

11. State Parks and Wildlife: A check with their office will provide information about any boats registered to the subject. Additionally, the records will provide any information related to any hunting or fishing license. The investigator should determine where the person purchased the license since this may point to a different part of the state where the subject may have a ranch.
12. Federal Civil and Bankruptcy Records: A review of the records will provide insight into the subject's financial responsibility. If a bankruptcy file since found, review the actual file as information about credit cards, hobbies that produce income, and all other debts and assets will be listed. A quick phone call can verify if the asset still belongs to the subject.

Once the public records listed above are researched, other more specific searches should be conducted. Several good web sites can be used and are good because they can conduct state-specific or nationwide searches. Some of these include:

1. www.knowx.com: This web site will allow you to search for stocks listed to the subject, real property, bankruptcies, boats, airplanes, lawsuits, judgments, liens, and much more. Often you will find information in other areas of the United States that involves the subject and no other information lead you there. Once you uncover a new location to search, you can zero in on specific records there.
2. www.iqdata.com: On this site you can check real property, social security numbers, drivers license, motor vehicles, and more.
3. www.dnb.com: Dunn and Bradstreet has financial and biographical information on companies and their officers.
4. www.pac-info.com: This web site provides access to property appraisal districts throughout the U.S.
5. www.kelmarpi.com or www.kellyriddle.com: These are my personal web sites that have more than 100 pages of search sites available by topic. Go to the site, click on record searches and then on "Internet on Demand." Click "yes" to download and you will have more than 100 pages of searches at your fingertips. If you want to search to see if a person has a boat, bankruptcy, criminal history, etc., this will give you a list of sites used to conduct the topic search.

The investigator should recognize that any information obtained through the Internet is an unverified source. It needs to be confirmed and correlated with other information. Also, most of the information on the Internet is not designed to be complete and should therefore be used in conjunction with public record sources.

Once these searches are done, you will have a pretty good idea of what you are dealing with. The next step is to conduct searches outside the U.S. and/or bank searches. These are very specialized searches and like everything else can be done legally or illegally. One legal way to determine a person's bank account is to create an alias name for yourself such as XYZ Marketing. Then, open a checking account under that name and send the person a check for $50 thanking them for trying your product or for entering your contest. Once cashed, it will go through their bank and back to yours where you can then look at the back and determine what bank they use and their account number. Foreign accounts are very difficult and require special contacts to access. These can be very misleading as well. For instance, in many offshore areas a person can become their own bank simply by posting $20,000 with the government. Many scams and hiding places can emerge from this.

Hidden asset investigations are very complicated and time consuming. However, they are not as complex as we are often lead to believe. The searches outlined herein are only guidelines. Many times the investigator will reveal information through these searches that require additional searches to be taken. Good luck and happy hunting!

Asset Checklist

By Col. (Retired) David W. Kuebler

- Checking Account Records: Subject, Spouse, Children, Deceased Relatives of both subject and spouse, Subject's first name and spouse's maiden name, Corporations, Partnerships, Joint Ventures, or Successions
    o   Multiple accounts with the same bank or branch
    o   Accounts where subject is listed as Secondary
    o   Family, Relatives, Friends, Groups and Organizations
    o   Custodial Accounts and or Trustee Accounts
    o   Reverse Side of Cancelled Cheques: Look for endorsements and routing
    o   Bank Statements with unusual activity: Deposits, Withdrawals, Electronic Direct Deposits, Electronic Direct Withdrawals; also Vendor Names & ATM activity
    o   Grandchildren Accounts: Subject listed as custodian or trustee
    o   Dummy Accounts using any of the criteria above with Social Security Numbers
- Savings Accounts with the same characteristics as the Checking Accounts
- Certificates of Deposits with the same characteristics as the Checking Accounts
- Money Market Accounts with the same characteristics as the Checking Accounts
- Safety Deposit boxes with the same characteristics as the Checking Account
- Security Facilities with Numbered Accounts
- Overseas Accounts and United States Customs Transaction Records
- Payroll Records and Employment Records
    o   Expense Account Records
    o   Medical Reimbursement Records
    o   Medical Savings Accounts
    o   Section 125 Medical Savings Account Salary Reduction Plan
    o   Severance Packages or Benefits offered due to termination
- Retirement Plans
    o   Pension Plans, Profit Sharing Plans, 401K, Salary Deferral Plans, Thrift Plans, Executive

Compensation Plans. Stock Option Plans, Provident Funds
- Life Insurance Policies with the same characteristics as the Checking Accounts
  o Whole Life Insurance
  o Interest Sensitive Insurance
  o Single Premium Whole Life or Lump Sum Interest Sensitive
  o Annuities—both Interest Sensitive and Fixed, Variable, Flexible, Single Deposit
  o Employer Sponsored Split Dollar Life Insurance
  o Employer Sponsored Executive Compensation / Bonus Plan
- Credit Unions with the same characteristics as the checking accounts
- Investments / Notes
  o Mutual Funds
  o Stocks
  o Bonds
  o Treasuries
  o Annuities
  o Hedge Funds / Investment Clubs
  o DRIPS—Dividend Reinvestment Plans
  o Bank Investment Instruments
- Federal and State Income Tax Records and Schedules for the last three years
- Part-Time Employment Income Deposit History
  o Cash Employment: Commissions, Tips, Per Diem
  o Unusual Non-reimbursable Expense items
- Employer Clubs and Civic Organizations with banking or credit account privileges
  o Private Clubs and Organizations with banking or credit account privileges
  o Hobby Clubs with banking or credit account privileges
- Property Real or Personal listed *nationwide* using the same criteria as checking
  o Titles, Bills of Sale, Deeds, Financial Arrangements
  o Real Estate, Land only, Time Share, Rental Property,
  o Automobiles listed in the names with the same characteristics as checking accounts
  o Automobile mileage records and maintenance records
  o Storage Space owned or being purchased
- Property Taxes and Sales Taxes

- Loans, Notes and Liens with the same characteristics as Checking Accounts
  - Mortgage
  - Secondary member to a loan
  - Co-Signed loans
  - Unsecured loans
  - Credit Card Convenience Checks
  - Personal loans with the same characteristics as Checking Accounts
  - Persons from whom money is owed
- Rental Property
  - Real
  - Storage Space
  - Garage
- Credit Card Activity
  - Direct Drafts from Credit Cards
  - Credit Balances on Credit Cards—Specifically large Overpayments
  - Credit Card Refund Checks
- Financial Statements and Credit Card Applications within the last 24 months on behalf of any lending institution, investment firm, brokerage firm, financial institution, credit card company, etc.
- Secretary of State Reports in all states in which any records of above reflect participation to include adjoining states.
- Accountant Information
- Hobbies
  - Collections
    - Guns, Coins, Stamps, Precious Metals, Baseball Cards, etc.
  - Licenses
    - Fishing, Hunting, Boating, Game Warden Record of Addresses, and Water Craft Registration
    - Watercraft and Aircraft Registration
- Last terminated employee, especially one of the opposite sex
- Former in-laws, except brother-in-law
- Mason Jars (Small hordes of money have been found in mason jars buried in the garden, bushes, and side yards with metal detectors)

Blue Knights Investigative Services specialize in: Surveillance, Domestic Investigations, Civil & Criminal Investigations, Background Investigations, Missing Persons, Polygraph, Asset Location and Appraisal, Insurance Investigations, Forensic Handwriting,

Fraud Detection, Protection, Escort and Bodyguard Service, Undercover, Corporate Verification, Workmen's Compensation, Fraud, and Theft.

Records Check: Employment, Credential, Education, Address, Telephone, Driving Record, Criminal, Civil, and Bankruptcy

Local, Statewide, National, International, & Internet Services Tailor Made To Your Needs.

Following the Money Trail
By Joan Earnshaw

When developing a paper trail, there are two general methods I usually use:

1) Analysis of bank account information, or
2) Tracking income without bank information through expenditures.

Bank Account Information
1) Deposits: How much and what is the source of these funds? Are these amounts in checks, cash, or traveler's checks?
2) Expenses: How much for each category? Look for transfers to other accounts that are made by check. You can later talk to anyone who received a check(s) to probe for further information.
3) Find what's missing: Are there checks for the basic necessities, such as food, house payments, utilities, etc.? If not, there's either another account or cash was used to pay those things.
4) Use spread sheet: Do your analysis in the form of a spreadsheet make it easier to look for patterns in income and expenses? Look also for names that appear repeatedly.

Tracking Without Bank Information
1) Find the statistical tables for the basic expenditures for an individual or appropriate size family. You know there had to be that much income at least. What are likely sources of income? Follow up by checking on sources that the person may use as a cover (welfare, loan from family member, inheritance, cash hoard, etc.) Remember that the person you are investigating is not going to admit to selling drugs or embezzling or having any kind of illegal income. Eliminating the excuses early on will assist you later.
2) Look for large expenditures.
3) Do surveillance to find out the vehicles owned or used, residence owned or rented, and other visible assets.
4) Check the county records for mortgages, judgments, and liens.
5) Check the tax assessor's office to see whose name the property is in and who is paying the taxes.
6) Check court records for divorces, prior convictions, and suits. Divorces especially contain a lot of financial information.

7) Check vehicle records for lien holders. Then check with the lien holder to see what the vehicle cost and if the person bought additional vehicles in his or her name or in other names.
8) Who are the person's family members and what is their lifestyle? Do they appear to own a lot of large ticket items?
9) Does the person belong to organizations that have a lot of social functions? Not only does this usually cause large expenditures on clothing, but also you may have some contacts within those organizations who can furnish information.
10) Remember to look for patterns in actions and relationships. This can be helpful since you can sometimes surveil the friends and family if the main person becomes aware you are surveiling him or her. You will probably find that they shop in the same stores and have the same habits. These same patterns will give you leads on your suspect.
11) Remember also that if the person is one of those who puts assets in someone else's name (such as mother's) or uses an alias, you will have to find some means of establishing that the suspect and the alias are the same person, whether you find a document, do a pretext, or third party testimony. If another family member's name is utilized, you can obtain an affidavit from that person (difficult at best) or you can establish that they do not have enough income to purchase the asset in their name. Then you can track the purchase back to the seller and obtain documentation as to who the real purchaser was.

My 26 years with the IRS provided me with some interesting experiences, not the least of which were finding money and assets of drug pushers, embezzlers, and the "Patriot" types. I still use a lot of the same methods It's not as easy for me to get bank records, but when I do, they provide a lot of information. I used to write a summons for account information, but now there has to be a subpoena issued to obtain it. However, I can now use databases outside IRS, so there are plusses and minuses.

It is very surprising to me that persons who try to hide their accounts in banks in out of the way locations will transfer funds from one account to another or one bank to another by check. Those same persons fail to realize that signature cards reveal additional accounts or safe deposit boxes. Items of deposit which are microfilmed checks, travelers checks, cash, etc., are also very good clues since they can lead to other names used, other accounts that funds are transferred from, and also provide information about customers' names. Customers whose names are obtained from the checks deposited can then be interviewed. Sometimes questioning one customer will lead to other bank accounts. When you find a customer who has paid by checks, look at the backs of all the canceled checks to see what account they were deposited in or if the checks were cashed. Sometimes receipts for expenditures will reveal how an item was paid for and can reveal the existence of a bank account.

Check at stores for charge accounts. It is helpful to consider all members of the immediate family. If a parent who does not have much of an income begins driving a fancy car, tracking down where it was purchased and how it was purchased will probably reveal that the child bought it for the parent.

It's also helpful to look at what does not show up in the account information. Everyone has to eat and have a place to live, so there should be checks for groceries and house payments, as well as other necessities. If there are not, then there is either another account (or more) or cash is used to pay for necessities. There may also be another person involved who is paying the basic expenditures.

How to find the bank accounts? Sometimes other law enforcement agencies will have that information and provide it. When I worked for IRS I used that source, but now I usually work without it. The lien holder name on a vehicle title is a good clue as to the subject's bank. His mortgage recorded with the county clerk's office is sometimes another clue to his bank. Occasionally I encountered persons who utilized other people's bank accounts, including the accounts of their family members or friends. It was a challenge to find all of these accounts.

Sometimes the best lead was from a bank teller or store clerk who was disgusted with the person hiding the money. Since most people are fairly consistent about depositing funds on certain days or at certain times of the day, they can often be observed doing so. What if there are not grounds to obtain a subpoena for account information? It's still possible to determine if there's an account. Call the bank, state that you're not asking for account information, but you have a check from person X for $__ and you want to know if it will be covered. You also have a right to ask if the account runs to four or five digits ($1000 is four digits—$10000 is five digits). This won't give you the balance, but it will give you the range of the balance that usually is in the account.

Sometimes one case will provide information on another case as money flows from one person or one business to another. Keeping copies of case reports can be helpful for that reason.

Some persons do not use bank accounts at all. Their income can be tracked by their expenditures. If they buy a vehicle for $25,000 cash, they must have had at least that much income. Their basic expenditures, i.e. food, housing, etc. can be computed from statistical tables, and this method is accepted in courts of law. Those tables are used unless the person chooses to provide either testimony or evidence for the expenditures, keeping in mind that large expenditures such as a vehicle, new house, jewelry, etc. are added to the table amount. Common large expenditures, i.e. "status symbols," are usually purchased from "in" stores. These very obvious items can be noted during surveillance and tracked to likely stores. The stores are usually cooperative in providing copies of receipts.

The most difficult to track are the barter cases. That is because the bartering is quite often done among a tight knit group of like-minded individuals who will not cooperate with you. The statistical tables were used a lot for those folks. Some surveillance was done to check for big ticket items, but the barter folks don't usually have big ticket items or a "spendy" lifestyle. The most helpful technique in this kind of investigation is to say, "If this were me, how would I handle my money?"

Enforcing Legal Judgments
By Brenda Collier, Esq.

While some of the enforcement procedures described in this article refer to the state of Texas, similar legal procedures are available in most states.

Getting a judgment is only the beginning in a collection process. The losing party rarely writes a check at the courthouse. The more usual scenario finds the winner needing to continue to use the judicial system to enforce a judgment.

The first step to enforcing a judgment is to protect the judgment from transfers of property by the debtor. One way to protect the judgment is to create a valid "judgment lien." This is done by properly recording and indexing an abstract of the judgment in each county where the judgment debtor owns, or might own, property.

The judgment lien "attaches" itself to all of the debtor's non-exempt property in the counties where it is recorded. The recorded judgment lien will also attach to exempt property automatically when the property ceases to be exempt, as well as to property the debtor later acquires.

Under Texas law individuals have certain property that is exempt from being seized for debts. However, corporations have no such exempt property. The judgment lien is good initially for ten years and can be renewed so that it stays in effect indefinitely.

Once the judgment is filed, you have a variety of remedies available to satisfy the judgment. But first you need to find assets of the debtor. This can be done through post-judgment questions directed to the debtor, interviews of the debtor under oath, or using databases or private investigators to search for assets. After determining what the debtor owns and where the property is located, you will be better equipped to analyze what additional steps should be taken in collection of the judgment. *See Christina Smiley's article "Collecting On Civil Judgments."*

One of the more common collection tools is a "writ of execution," which is a judicial order directing the enforcement of a judgment. The writ instructs a court officer (sheriff or constable) to seize the debtor's non-exempt property, sell it at auction, and deliver the proceeds to you.

Texas law directs that seizure will first be upon property designated by the debtor when the writ is served. However, if the debtor fails to designate specific property to be seized, or if the court officer believes the designated property will not sell for enough money to satisfy the execution and costs of sale, the officer will require the debtor to designate additional property or simply seize any of the debtor's non-exempt property.

If you find cash or learn that the debtor has funds or property owed to it by a third party, you might choose to exercise the remedy known as "garnishment."

You can obtain a garnishment judgment, which orders the third party (might be a bank) to pay the funds or property to you rather than to the debtor. A garnishment order is issued by the Court without prior notice to the debtor so that the debtor will not have an opportunity to move the liquid assets. When the order is served on the third party, the debtor's account is temporarily frozen. To obtain a garnishment order, the judgment holder must tell the court that as far as she knows, the debtor does not have enough property in his possession in Texas to satisfy the judgment.

There are risks to filing a garnishment action. For example, if the third party is not holding money or property owed to the debtor, you may be responsible for the payment of the third party's attorney's fees to respond to the garnishment.

Garnishment is most frequently used to seize cash or liquid assets. It usually gets the debtor's attention quickly.

Another commonly used post-judgment remedy is the "turnover order," so named because it literally orders the debtor to turn all non-exempt property over to the judgment holder. This remedy allows a judgment holder to spread a wide net to catch all available assets when the debtor's property cannot readily be attached or seized by an ordinary legal process. When all your efforts to locate the debtor's assets fail, you may want to seek a "turnover order." This remedy is usually used when no other means to satisfy the judgment are available. However, it does not require that a judgment holder exercise all other post-judgment remedies before seeking the order.

Methods of turning that judgment into cash are complicated and each comes with its own cost and risk. You will want to evaluate each available post-judgment remedy on a risk-benefit basis before proceeding.

The information provided in this article does not represent legal advice. If and when you face a specific legal situation, you should conduct independent inquiries with legal professionals to determine what your legal rights may be.

Nabbing Assets Before They're Eaten Up
By Edmund J. Pankau

You have gone through the full legal process, won your client's case, and obtained a substantial judgment. All you have to do is levy the assets of the other party and reap the rewards. But is it that easy?

You've perfected your judgment, the opposing party refuses to pay the debt, and the sheriff or constable cannot find sufficient assets to satisfy the judgment. What happens next?

Most attorneys first try to depose the other parties in a post judgment discovery action to determine their net worth and then locate the source of assets. You should require the opposing parties to present their financial statements, tax returns, and bank statements. Also, question subjects about their income and personal and business assets.

During this post-judgment discovery, many attorneys find that the opposing parties have already hidden their assets. Subjects may have either taken the money offshore or transferred it to other entities, such as their children's or family trusts. Or they may have hidden it using another name, such as the spouse's maiden name. Assets will be there nine times out of ten.

In twenty years of investigation I have heard almost every story about how people have lost their assets and now can't pay the judgment. Typical stories include: "I went to Las Vegas (or Atlantic City) and lost it all gambling." "My wife and I used the money to live on and spent it all." "My bookkeeper or accountant stole the money." "I lost it all on bad business deals trying to raise the money to pay this judgment."

You may have heard similar stories and found that the parties could not document their claims. Let me offer a few suggestions that may show you where the money has really gone.

- Check to see if the subjects made large payments on their home mortgages during the period of your lawsuit, especially in the last year of the lawsuit. Many people try to hide their assets by paying on their mortgage and adding to the equity of their homes, which in Texas they believe to be "bullet-proof" because of the homestead law.

- Examine the payment of their universal life or whole life insurance policies. A prepayment can accrue interest just like a savings account and doesn't show up on financial records except inside the insurance policy itself.
- Look for savings bond purchases, either in the subject's names, their children's names, or the spouse's maiden name. Until recently, these transactions were not registered centrally and were a favorite purchase of money launderers and drug dealers.
- Look for cashier's check purchases in the bank accounts of your party. These checks can be purchased and tucked away for the future just like cash.

When you bring opposing parties to depositions, instruct them to bring the kind of documents that will help you and your investigator trace their financial history. This is one of the most overlooked areas of discovery because attorneys frequently don't understand the process of investigation.

Where a person went and who they spoke to or dealt with is often more important than knowing about their business and where their money was when those last financial statements and tax returns were filed. Sometimes the most crucial evidence can be found by tracing such activities. The timing of these activities may prove that financial transactions were made to protect personal assets.

The following key documents may reveal a person's hidden assets. They may also help prove the intent of hiding these assets from the court.

Passport
When you subpoena parties, request their passports in your motion to produce. The entry and exit visa stamps will disclose trips to places such as Switzerland, the Cayman Islands, the Bahamas, Isle of Man, Netherlands Antilles, and other locations. These trips may have been made to hide money.

By documenting financial withdrawals from bank accounts and timing them with trips to foreign countries, you can often discover offshore fund transfers.

Telephone Records
Many smart attorneys request business or personal phone records, but they don't think to include their subject's mobile or car phone records. Remember that standard telephone records only record

long distance calls. Mobile phone records record all calls, local and long-distance, for billing purposes.

If you want to find an undisclosed business partner or a "significant other" in a relationship, try subpoenaing the car phone records. Compare the numbers and names on the bills against the people who are known to your clients. I promise that you will find some interesting information.

Credit Card Statements
These records document out-of-town travel and often name people with whom your subject has had dinner or done business—people who may well be pertinent to your investigation. These records also indicate what hotel the subject has stayed in.

Hotel Records
Hotels now invoice not only for the room, meals, and drinks but also for the long-distance phone service. Calls are often made to those people the subject wouldn't call at home, such as an out-of-town banker, business associate, or third party.

Credit Reporting Agencies
The trail of credit purchases follows us around the world. Like expense accounts, a credit report lists where we eat, stay, and make purchases. It also contains a record of other business entities and banks that inquire about our financial status. Credit card inquiries are one of the better ways to locate undisclosed bank accounts, insurance policies, and other major purchases.

Airline Travel
When questioning parties about their business activities and assets, always ask for the name of their travel agencies. Once you have this information, another subpoena can be sent to the travel agency to document all the trips made by the subject.

Telex
When investigating a company's assets, always consider the company's telex records to find foreign business entities, foreign bank accounts, and other offshore activities. The monthly telex log or bill will point you in the right direction.

Overnight Packages
Almost all of us use Federal Express or a similar carrier to deliver our valuable mail and packages around the world. Examining the monthly bills for a company's overnight packages gives a clear idea of the cities and countries that it is doing business in and will always add more information to the discovery process.

All of the records listed are easy to obtain, particularly under a motion to produce. If you find that the opposing party is unwilling or unable to gather these documents, you should think about subpoenaing these records directly from the sources that produced them.

In many cases once these records have been requested, obstinate parties suddenly become much more amenable to resolving financial issues and settling their judgments. If they have anything to hide, they would much rather settle with you, an obviously experienced investigator than produce these records publicly for other creditors to find as well.

The next time you go into post judgment discovery, or if you want to find the financial worth and assets of parties earlier on in a case, consider these ideas and resources for your legal strategy.

Chapter Six

Information Gathering

Public Records in Depth
By Joseph J. Culligan

The county is the depository of many records that can be searched by hand. There are hundreds of different records that may assist you in finding someone or doing a background check.

Alarm Registrations

There is an investigative technique that I frequently use that surprises everyone. It is the alarm registration.

When someone uses a company—Brinks, for instance—to be their alarm company, that person must complete a registration form for Brinks. Then the person also has to fill out an application to the local government. The Brinks application is not public information, but the one made out to the local government is.

The reason you fill out an application for the local government is so that the government will have a way to contact you if your alarm goes off more than three times in a year, because if that happens you will be fined (that is the rule of thumb for most governments). The application you fill out is a public document because there is a relationship between you and the government. Since they can fine or tax you, that record must be open to public inspection.

I have always used public records to the greatest extent possible because of the consistency that they provide my investigations. To depend heavily on a data base provider who supplies records out of the mainstream has many pitfalls and not just legal ones. Since they are simply not dependable, you do not want to have your entire case hinging on information from a broker.

An alarm registration of an individual or company can be looked up by name or address. The record is usually found at the communications department that handles 911 calls for the jurisdiction or municipality that you are searching. For instance, in Dade County, Florida, it would be the Dade County Police and Fire Communications Bureau, but in Miami where they have their own 911 system, it would be with them.

The typical application will contain your name and telephone number. If it is unlisted, it still has to be listed on the application because they need to contact you. No one gives a wrong telephone number to the police department on an application like this, so you can be assured that the telephone number is good. This is a great way to get unlisted telephone numbers.

The next information listed on the application is place of employment and its telephone number. This is a good way to find out where someone works.

The next bit of information listed—the next of kin or someone who is allowed to go into the house or give authorization to the police to go into the house—is what is most important to me when finding someone who is difficult to locate. The person who is listed is someone, as you can imagine, who is very close to the subject. No other document that is so readily available gives you a person that may know where your subject is.

There are two distinct files I ask for. First, I ask for a copy of the original application because the current information may list an employer from years ago who is not on the present printout. Also, a next of kin may be listed on the original application who is not currently listed on the computer screen. Remember this is important because the person listed on an application from many years ago and who is not listed now may have an ax to grind against your missing subject and could possibly give you much valuable information.

When you pass by the residence of your subject, you can see if they have a sign that says Brinks or some other alarm company. I do not make the observation of a sign the criteria as to whether or not someone has an alarm. I simply get a list of all the addresses where my subject has lived at and run those addresses. Old addresses that the subject does not live at are fine to run because you still want old applications since they can provide much information such as former employers and next of kin.

Typically you can contact the local police department and fire communications department to retrieve a copy of the orginal application and, most importantly, a printout of what is shown on the computer screen, which of course gives you the most current information.

In a major political case from a few years ago I was able to get the unlisted numbers of several people. I had to revisit the story again not long ago, and when I requested the information I was refused. I called the Commander of the Communications Department who knew who I was. He said that he had new applications printed that had a box that can be checked by an applicant as to whether the applicant will allow the information to be released if requested by a third party. I told the commander he could not do that. He said he could.

I contacted the county attorney and said that no department had the right unilaterally and arbitrarily to make a public record not public at the request of a citizen. I explained that the tax collector could have the same box on tax records so people could keep their property tax records confidential. Or a person could check a box on the divorce records and keep them secret. The county attorney understood. The public is not the party to decide what is public information.

The commander was told by the county attorney that he had no authority to close up public records at the request of a party and that he was to open up immediately the alarm registration records to public inspection for me and anyone else.

Pet Licenses
Pet licenses are a record that I always ask for in an investigation. If the subject owns a pet, I am sure I will be able to find him or her because I know that anyone who owns an animal who is dear to her, will have the pet registered.

Check the counties in the state you think your subject may be in. Dog licenses are a matter of public record, and they are, listed by the owner's name and not the animal's. Every person who cares as much about their dog as most people do will not only keep their dog licensed but will give the correct information such as home address and telephone on the application so that in the event the dog is missing they can be contacted. No one gives bad information on a license for a pet if they ever want the animal reunited with them in the event of his disappearance.

Computer Records vs. Hands On
Recently I gave an interview to a newspaper in Arizona. The reporter said he wanted to do the article about how the computer has changed the private investigation business.

He asked me what type of Private Investigator (PI) I was. The kind that used the computer or the kind that did it the old-fashioned way? I told him that it was not a question of either/or, but that a PI has to use new and old techniques in combination to enhance each other.

For instance, if I was doing a very detailed background check on someone, I could not just run computer checks. I would ask the Department of Motor Vehicles (DMV) for the "body history" or "body file" of all vehicles ever owned by the subject for a small fee *(check if your state law still allows this record to be released).* These requests will yield me copies of all the transactions regarding a vehicle from the day the car left the manufacturer. Everyone has been a buyer and seller of a vehicle, and small talk a person makes with somebody they did the transaction with can uncover important information. While buying or selling a car, a person says things, maybe not very important, that are things they would not bother to mention to anyone else. That is because they are just killing time and just want to make small talk during the transaction, or they just want to make the other person feel comfortable. I go back to interview the people the subject sold a vehicle to or bought one from.

I had a case where the subject mentioned that he had to pick up his girlfriend at the bank she worked at after 5:00 PM and that he was getting a cashier's check at that time for a car he was purchasing. I discovered that the subject was using this relationship, which had occurred six years before, as a way to wire funds without giving the required information that is needed for funds over $10,000.00. He was not dating the bank employee any longer when I interviewed her. She had kept a copy of all the transactions she had done for her "ex" friend. This led me to dormant accounts in this country as well as offshore, and I simply followed the paper trail from there to find out where the money finally went. This case was not a marital case, but rather a corporate investigation because the subject was stated to be named as a CEO of a major brokerage house.

My client was given the information about how the subject hid money that he was skimming off of deceased client's accounts. The subject would take control of accounts of deceased persons when heirs failed to come forward after about two years. Since he was an executive officer of the brokerage house, he was able to carry out this scam for 52 accounts that were identified with the help of a branch manager confederate. Because of a review of a vehicle transaction from six years before an invaluable piece of information was discovered. This could not have been done with a computer.

Military Records
Many times when you do a background check, there will be several different reports that you will need to gather to make into one report. I always compare the various reports to make sure that all the dates fit—the same date of birth, social security number, etc.

Throughout life people have many difficulties, and some may be inclined to change parts of their background to conceal things from others. I always order military records since they inevitably prove to be a good source of "real" information on a subject. A person generates military records at a younger age than most other records, and military records are frozen in time so you can be assured that they have not been changed or deleted in any way.

The background checks I do are very extensive, and my clients are usually willing to spend whatever it takes. Here are some of the ways I use military records.

When you run a normal credit header report on someone, it shows past addresses. The past addresses in the credit report give you a starting point to find people in someone's past in order to question them. Or it will give you different geographic areas to search for civil records, such as lawsuits, divorces, the ownership of property, etc. But if a person was in the military, the header report shows addresses such as APO and FPO. You have more information available for your search than you would have had if that person had been a civilian.

Part of the records you receive from the military will contain Past Duty Assignments. These are very important. If your subject had been stationed in San Diego, you will be able to search the public records for criminal or civil actions in that area. Perhaps there was a paternity suit or a marriage. Or perhaps a divorce after the marriage. Also past military assignments will give me a clear picture if the subject has been trained in explosives or is an expert in artillery, including guns. I need to tell the client that the subject has a background that has a generous sprinkling of skills that she or he may need to know so that they can protect themselves or even obtain a restraining order. Remember restraining orders are not just for battered wives. Many businesses are forced to get peace warrants against a former employee when they create disturbances and have the potential to create more disturbances.

Also, do not overlook the awards and decorations that the subject received. These will give you a clearer picture of where the subject was when he was in the service. Some special decorations are for special forces, such as the Army Rangers or Navy Seals. Also, many intelligence officers are recruited from the military.

Now you can see that if you had done just a standard credit header report, you would not have had an address for past or present military personnel unless the subject had lived off base. That's why military records can be very essential for most background checks.

Driving Records

As we all know, driver licenses are a great source of information. Each state varies on what information is contained in a driving record, but let us take the State of Florida driving record as an example. You can get a person's Social Security Number, date of birth, list of accidents, dates and locations of traffic tickets, eye color, hair color, weight, height, home address, the date the subject was first licensed in the state, in what state the subject had a driving record previously, and more.

Very seldom is it the case where I run a driver's license and get a good address for a subject. To get the subject's address I usually run motor vehicle registrations because unlike driver's licenses, which may not have to be renewed for sometimes up to six years, a vehicle has to be re-registered every year.

I use the driver's record to look at accidents. If I see an accident, I get a copy of the accident report. This record is one of the best investments in time that you can make. I look to see who is at fault. That is important because if your subject was at fault he may be getting sued, in which case you would look to see if a lawsuit has been filed. Go to the file and look for the process server information. If good service had been completed, then you will now have a good address for your subject. If the lawsuit was many years ago, then you may have an address you never knew about. If your subject was not at fault and it looks like there was injury or a lot of property damage, then your subject may have filed suit. Your subject's address on the accident report is the address from his driver's license. That may not be a good address. But you can be assured that if your subject has filed a lawsuit against the other driver, a good current address will be contained in the lawsuit filed. Your subject wants to collect monies and if he gave a bad address in the lawsuit file, then his efforts to collect could be thwarted.

Please do not overlook the previous state of licensure. You can write to that state for the driver's license, and, again, you may find a treasure trove of information.

Now the bad news. It is common knowledge that the downfall of the accessibility of driver's records started when Robert Bardo, a stalker, hired a private investigator to find a young actress for him in California. The private investigator simply ran the driving record of the actress, got her address from the license, and then gave it to Bardo. Robert Bardo went to the actresses' home and when she opened the door, shot her dead. That is all that was needed for California to restrict access to driving records.

California has an excellent service whereby they will forward a letter to someone if a person writes to the DMV and asks for a letter to be forwarded to a person. It works the same way as the Social Security Administration's letter forwarding service.

In 1994 President Bill Clinton ordered the Department of Transportation to have all the states be more responsible about the release of information from driving records. If they did not comply, then the funds for highway improvements would be withheld.

By 1997 all states submitted their plan to the Department of Transportation concerning what their plan would be to protect the public from unwarranted invasion of privacy. Remember that the states were generating revenue from the sale of driving records and were reluctant to comply, but the funds for highway improvements was an overriding factor.

Note: If this service does not currently exist, petition your state to implement California's letter forwarding service through its Driver's License Division.

Nannies

There has been lately a proliferation of companies that offer background checks on nannies. That is for good reason, as you have seen the destructive examples of what a child care provider can do when left unchecked. The results of what the nanny cams have taped that you have seen on the evening news is something that none of us can forget.

Even though I am unable to tell you who I have done background checks for regarding their nannies I want to let you know what I do that may be a little different insofar as what I supply to the client.

The nannies who are hired by celebrities come from agencies. These agencies do a cursory check of the application that is submitted. Some of the better firms check the criminal background, credit rating (so the client can feel that they are not hiring someone who is desperate for money), and driving records.

What I do is make sure that all the time is accounted from when the nanny left her parents house at age 18 years old or younger. I want an address for each day that this person has lived, and I do not care if I go back 40 years. I want to know who lived at that address, the reason why the nanny left that address, what the police and fire/rescue reports say for what has happened at that address (when available), and I want to know who owned the property, etc. Also, of great importance is the fact that if the child is ever taken, I will then have addresses and past acquaintances and family members to go to immediately. Many times the FBI and police only have current information about the nanny, but there is nothing like also having the past addresses readily available. Just where do you think nannies take kids? Many times to places they feel other people would not know about and where they will feel comfortable.

The most important thing I do is a background check on former and present persons who have been involved in relationships with the nanny. As I tell every famous client: when the boyfriend breaks up with the nanny, do you think that he will do it calmly at some remote location, or will he do it at your mansion and make a scene in front of your child when you are not home? Remember, the nanny lives at the home of the famous person, so that's where the breakup will take place.

Also, I need to know the finances of the boyfriend. Will he be poking around the mansion looking for things to steal? Will he try to find something incriminating that can be sold to the tabloids? I did a background check for one of the television hosts I have worked with for years. The former husband of the nanny had been a lawyer, but he had some legal problems and was now selling newspapers for the Cleveland Plain Dealer. I told the client that just because of this, the nanny had to be disqualified. The former husband was a man who had been broken and could therefore possibly see the nanny's employment as a source of money whether through a scheme or outright theft.

If you are thinking I might be a little too cautious, then I say to you why should I believe that an attorney who has been busted for fraud is going to be reformed and will not be jealous of his former spouse who is doing so well in the company of famous people?

There is so much to background checks that must be done, but rarely is. I know that resources are a major consideration. Does the client want to pay for all this extra digging out of information on peripheral characters when nothing may come out of it?

Record Accuracy

Many times when I run reports from the various database companies, I come across errors. When doing the background checks of many people for the media that was necessary during the post presidential turmoil of late, I saw many mistakes in the reports I got back that could have caused serious problems insofar as liability and litigation.

I ran several reports on a particular person who was the spouse of one of the principals. The most current driver's license report I got back showed a phony Social Security Number. Before I would release this information to my client, who was to go on the air and say that his spouse had committed perjury on the driver's license application, I called the database company. They said that the information on driver's licenses was bought from another company. I asked specifically if the information that was on the driver's records was from the driver's license bureau directly. They said they did not know, but they would check with their vendor. The next day the database company I used called me back to say that the vendor said that they sometimes put other information on the driver's license records from other sources. So there you have it. Some of these companies sweeten their records and commingle information from other public records sources, such as marriage records, UCC statements, divorce records, and many other types of sources.

The very important lesson to be learned here is that unless you go down and eyeball the record itself, be cautious about telling your client that something is a fact. Remember that it is important not to accept the results of your search at face value.

Arrest Records

I recently went to the Miami-Dade Police Department to see if a particular person had ever been arrested in the county. I gave Clerk Rondon the information. He said that it would take a few minutes. The computer came back with the information that the subject had been arrested. I told the clerk to make a photocopy of the printout. He said he could not give me the actual printout, but he would transcribe the arrest information on paper with the police department's logo. I said fine.

He gave me a completed letter that only showed that there was a record of an arrest and nothing else. I told the clerk that he was required to give me all the information that was on the printout. He said no, that it was their policy to just confirm that there was an arrest.

I left the records department and went to my office where I called and asked for the head of the records department. She was not in, so I left my telephone number. The records supervisor called back and asked whom I represented. I did not tell her I was a private investigator and did not tell her I worked for the media and was working a story. She said the contract they had with the contractor who supplied the computers stated that the public could not have arrest information and that only law enforcement, attorneys, and certain others were allowed to receive the information. I told her that she was in error and that the public records law overruled any agreement that the police department might have entered into with the computer company that gathered the information from other states and the Federal government. I thanked her for her time and asked that she double-check with the computer company.

I immediately called the County Manager's Office of Miami-Dade County. I asked for the Assistant County Manager in charge of overseeing the Police Department. His name was Paul Philip, and he was the former head of the FBI in Miami. He had been hired the previous year to oversee the county police department that had a very poor record of being responsive and responsible to the public. In large counties there are assistant county managers who oversee every department, such as Parks and Recreation, Property Appraisal, etc.

Assistant County Manager Philip did not call me back directly, but within the hour I had a call from a police official that was in complete agreement with me that arrest records are public information. He also said that Clerk Rondon was a probationary employee that had only been on the job for two months.

Shortly after that call the records supervisor called me back and said that there had been an error in the policy of limiting arrest records to the public. She said that the public was, indeed, entitled to the arrest information.

The aforementioned is just a routine day at the office. Before I wrote my books over a decade ago, the great majority of record departments in the United States refused to release public records. As more and more people went to record departments for records over the past decade, the offices have complied with requests more readily. If a public record is refused then you should take the time to contact the appropriate authorities. The refusal you encounter may be just the result of a new employee's misinterpretation of a policy, or the unilateral and arbitrary decision of a records supervisor, or the combination of both as in the above case. You must understand that the unresponsiveness of government officials that was prevalent years ago is a thing of the past because they are now aware that custodians of the records are subject to civil and criminal penalties.

Modified excerpt from my book, *You Can Find Anybody.*
Voter's Registration
Voter's registration information is available upon written request. Write to the county you believe your subject may have been or is registered in.

It is easy to ascertain a subject's date of birth and home address. In one case the only information I had was that at one time the subject named Lub lived in Baldwin, New York. Since I was not supplied a first name, I wrote to the Board of Elections of the county of jurisdiction for Baldwin. I sent a nominal fee of $3.00

since this is the average cost of requesting voter's registration records.

I requested that I be sent the voter's registration information of every person named Lub in Baldwin. In this case I was fortunate that my request was for an unusual name. I learned that the subject, Lub, did not live in Baldwin any longer and had moved in 1985. The subject's new address was 140 Larch Street, Wantagh, New York. The response I received also included the subject's date of birth. Since I was aware that my subject had been in a serious accident on a certain date and at a certain location, I asked for the subject's driving record.

To be able to write one letter and be provided with a full name for the subject makes this one of the best searching techniques.

Joseph Culligan is a licensed private investigator and the author of *You Can Find Anybody!*, *When In Doubt, Check Him Out*, and *Manhunt: The Book*. He is a Hall of Fame member of the National Association of Investigative Specialists, works mainly for the national media, attorneys, and government. Mr. Culligan, as well as his cases, have been featured on shows such as Montel, Maury Povich, CNN, Hard Copy, Unsolved Mysteries, and many more.

Note: A fully automated fill-in-the-blanks Public Records Request letter generator can be found at the Student Press Law Center on the internet at www.splc.org/ltr_sample.html. Click "State Open Records Law." Available by individual state with cites for each state statute.

Information Databases
By T. A. Brown

Information databases are databases that contain much of the information that help to compile a background check on employees and future employees. Some of them are expensive, while others are not, providing a "no hit, no fee" service. Many of the databases are available to businesses while all of them are available to private investigators and require a Private Investigators license number to access the database. A couple of them are available to the general public. Almost all require a credit card for the charges.

The top five databases are listed in the order of their most popular use and preference, based on a two-year study and survey by thousands of PI's who do background checks on a daily basis. The rest of the databases mentioned are listed randomly in no particular order. Most of these databases can provide flexible searches using limited information, but keep in mind that no one database can do it all. Many PI's use at least two or three for crosschecking and to obtain the most information. Sometimes databases will produce different results. One database may turn up "no match," while another will provide information. That is why most PI's use at least two databases for all their employee background checking. Here are the top five databases:

1. LocatePlus is a nationwide service. Least expensive. Provides some of the most accurate and up-to-date information out of all the databases. Initial searches are only $1.00 each on a no hit, no fee basis. Social Security searches are also $1.00. More information can be obtained for as little as 50 cents more. Provides a significant amount of information but can sometimes be about three months behind current information. They offer 24 hours, 7 days a week access. There are no sign-up fees, no connection fees, and no monthly fees. However, you will need to submit a copy of your departmental letterhead. Lists more neighbors and returns more date of births. Said to be well worth the money. 888-746-3463 www.LocatePlus.com

2. Merlin and Flat Rate are about the same and obtain their information from the same source and the annual fee is the same. They are among the most expensive with an annual fee of $1400, but more than worth the price, especially if you will be using them a lot. You can make arrangements for payments. Merlin does offer more search options. Price includes all the searches of any type as often as you need

them, and it includes some other "pay" searches that you may want to use occasionally. Also has a free "Ultimate Weapon" search that shows you where you will find hits before you do the run. You can save a lot of time and work because you can search each name several ways, with common misspellings, a check of relatives, etc. without worrying about the cost. Be aware, however, that criminal records are quite often misleading. For example, if someone has a simple traffic violation, it will show up as a criminal record in Merlin. Otherwise, Merlin is said by users to provide the best information in the Western U.S. Call them and ask for a free trial. 800-367-6646 www.merlindata.com

3. Accurint/IRBSearch is new to the business but said to be better than Flatrate by PI's. Accurint and IRBSearch use the same information resources although each database focuses their services for specific industries. Accurint services collections, law enforcement, and law firms. 1-888-332-8244 www.Accurint.com.

IRBSearch provides web access to over 20 billion records, mainly to the private investigative industry and collections. There are no service charges, or minimum search limits per month, and it is a pay per use search. No searches over $5.00 with many searches starting at 25 cents. They go back up to 30 years using more than just credit headers and give more detailed information. There is also an "age range" feature. It has the added features of relatives and associates and a "blue check mark" that shows if a phone number is confirmed as being active. Many sources are updated daily, but not all. Free trial provided. 1-800-447-2112 www.irbsearch.com

4. IQData & Loc8Fast are pretty much the same, since they operate out of the same location and use the same searches. They are also affiliated with Conficheck and U.S. Public Information. All information comes from the same databanks. The only difference is the price. Loc8Fast is much more reasonably priced. Also, IQData can be "down" for long periods of time. www.loc8fast.com

5. PublicData is open to anyone and very inexpensive at 10 cents per hit. A very useful tool. Great information and starting point, but information should be verified since it is somewhat outdated by up to a year or more. Offshore company to avoid Texas privacy laws. 972-869-2471 ext 111 www.publicdata.com

Several Other Databases

- Add123. Florida based only. Current information. www.add123.com
- AutotrakXP. Very expensive and allegedly has some inaccurate information. Good for the southern U.S. atxp.dbt-online.com
- Ameridex. Rated high for military locates. www.kadima.com
- AT&T. Directory for locating Businesses. Free. Highly recommended. www.att.com/directory
- Confi-chek. Nationwide information provider. 800-821-7404. www.confi-check.com
- ChoicePoint. Formerly CDB Infoteck. Pricing information available upon application. www.cdb.com/public
- Data-trac. No monthly fees, no sign-up fees, no per minute fees. Search fees start at $1. Used by judgment collectors. www.data-trac.com
- DCS. Inexpensive. Dallas Information Systems. 972-422-3621, 800-299-3647. www.dnis.com
- DBT. Very expensive. According to many PI's, allegedly has poor service. Information is about six months old, but their national profiles are said to be good. $25 per month usage fee, used or not, and $2.50 per minute online charge, plus other charges. Said to be best at locating relatives' information. www.dbtonline.com
- Infobel. International database. www.infobel.com
- Fosson. Free Public Records. State-by-state listings. www.fosson.com
- Netronline. Public records. www.netronline.com/public_records.htm
- Savvy Data. No monthly fees and quick results. www.savvydata.com
- Search Systems. Many free databases. www.pac-info.com
- XpertSearch Inc. Does county criminal records and charges from $3 to $8 per county nationwide. 904-447-7841. www.XpertSearch.com
- PallTech. Low-cost. Five billion records in their databases at $1 per search. No hit, no fee. www.pallorium.com/PallTech.html
- Knowx.com. Good basic database resource. Does not report county where record was filed. www.knowx.com
- Masterfiles. Low rates. www.masterfiles.com
- US Find. Nationwide people search for $7. www.usfind.com
- US Tracers. www.ustracers.com

- QuickInfo.net. Specializes in flat-fee pricing and is from the same company as FlatRate. However, the information is state-by-state and governmental and business networked. www.QuickInfo.net
- Pacer. Fairly inexpensive at 60 cents per minute. Extra software required. National Index for Federal Court information and detailed information for the Southern District of Texas. pacer.psc.uscourts.gov
- Source Resources. Has criminal, business, corporate, and other records. No sign-up fee. www.sourceresources.com
- iQ411. Real time telephone listings updated daily. No hit, no fee. www.qsent.com
- Intellicorp. Formally US Public Records). Good accurate criminal and civil searches at reasonable prices. Cheaper and faster then DBT. www.intellicorp.net/info/login.htm
- Lexis-Nexis. Provides authoritative legal, public records and business information, etc. www.lexis-nexis.com
- Public Records Resources. Free and subscription resources. www.publicrecordsources.com
- U.S. Crime Search. Public record searches in real time. Open to the public. Criminal records are not always complete. A county-by-county search is the only true way to verify those records as not all law-enforcement agencies turn in their reports. www.uscrimesearch.com
- United States Interlink. Very high ratings for up-to-date information. No monthly fees and searches are priced as low as $5 - $15 per search. 888-682-7764. www.usinterlink.com

Miscellaneous Tips
By T. A. Brown

The World Wide Web is a fickle and ever changing forum, so there is a possibility that some of these links may no longer exist, but to the best of my knowledge, they are still active.

Alarm Trips
The Protector Plus Voice Dialer Security System (DS7000) is a wireless alarm system for your business or home and has an infrared motion detector. It's connected to an audible alarm and a phone dialer which will dial up to 4 numbers consecutively until someone presses "0" on their phone to listen in and disarm the system. It includes one base console, two door/window sensors, a motion detector, lamp module, Keychain remote, and security/ home control remote. You can get it at www.x10.com (X10) for $99 at www.x10.com/products/x10_ds7000.htm.

America Online (AOL) Contact Information
General information: 703-265-1000
Screen name information: 888-265-8004
Members who want to cancel can do so over the phone, online, by fax, or by mail. Ph: 1-888-265-8008. Online at Keyword: Cancel. Fax: 1-801-622-7969. Address: P.O Box 1600, Ogden UT 84402-9926.

To subpoena records from them, they will not record out of state subpoenas. Litigation needs to be filed in Federal Court or in Virginia where the main office is located. Subpoenas must be personally served at: AOL Custodian of Records, 22000 AOL Way, Dulles, VA 20166. The account holder will have the opportunity to quash your request, but if they don't you should obtain your information within two weeks.

Bankruptcy Search
Black Book Online offers free nationwide bankruptcy searches courtesy of data provider, www.loc8fast.com. The searches can be run by debtor name or Social Security Number. Information returned includes city and date of bankruptcy filing. Fees will apply to obtain detailed information at: www.crimetime.com

Bomb Threats and Physical Security Planning
Do not allow a bomb incident to catch you by surprise. The following website gives a detailed list of what to do if you think your business may have a bomb planted on the premises, including

illustrations of how to protect employees and where to search for the bomb. This is a wonderful source of knowledge. www.bombdetection.com/btapsp.shtml#Bombs

A prototype Cryo3 is the first hand held radiation detector, not only allowing security personnel to quickly check for radiation anywhere but also to determine the exact amount and type. For visual as well as more detailed information, go to www.popsci.com/popsci/science/article/0%2C12543%2C259452%2C00.html

Bug Detection

Almost all bugging devices work on an FM frequency because they are simply wireless transmitters. If you think a room is bugged, have a radio tuned to an AM station in the room with the volume turned up fairly loud. Then use another radio tuned to an FM station and start turning the dial very slowly, starting at the bottom of the numbers. If you hear the station that the radio was tuned into from the AM station on your FM radio, then you know you are bugged.

If you hear a squeal or feedback on the radio, that means you are getting closer to the bug. The more squealing or feedback, the closer you are getting. This tip was given by Joseph Culligan.

Recommended TSCM firms: www.tscm.com/goldlist.html

Business Locating and Information Resources

Go online and register for free information that will help you locate a business or person at: www.freeERISA.com. The following information is available in detail at this website.

- ERISA Form 5500 Filings
- IRS Form 5310
- EIN Finder
- Public Pension Funds
- Top Hat Plans
- Provider/Client Database
- SEC Filings
- Tax Exempt Organizations

For other great business information such as employee numbers, wage reporting, business law and regs, etc. go to: www.firstgov.gov

For directory of boards of professional and occupational licensure in North America, go to: www.clearhq.org/boards.htm

Business reference desk: www.libraryspot.com/businessinfo.htm

Canadian Resources
One of the most complete listings online of Canadian Resources, including bankruptcy records, business contacts directory listings, banks, etc., and has worldwide records as well. Search Systems: www.searchsystems.net

Census Statistics
"State and County QuickFacts" permits users to browse quickly (via pull-down menus) for national, state, and county level statistics. www.census.gov/qfd

Collection and Credit Forms (free)
Sampling: Acceptance Of Collection Claim; Application for Open Account Credit For Stocking Dealers; Bad Check Notice; Bad Debt Write-Off Worksheet; Credit Information Request; Dispute Of Information On Credit Report; Notice to Correct Credit; Notice to Stop Credit Charge; Promissory Note - Demand Joint and Several Liability; Request for Credit Interchange; Revocation of Guaranty. www.alllaw.com/forms/Credit_and_Collection

Confidential E-Mail Program
If you want to send confidential information securely through the web between your company and your client, get a free easy-to-use secure email program from www.hushmail.com

PC Magazine Editors Choice for Secure Business Email is found at www.zixit.com/index.htm

Copyright Search Tool
Discovering who owns what in terms of intellectual property can be an overwhelming task for authors and others trying to secure permission to use copyrighted material. The U.S. Copyright Office has now developed a simple tool to search for ownership of rights in three major categories—serials, documents, and registered works, such as films, music, software, and works of art. The search covers works registered since 1978, with some 13 million search terms available. www.loc.gov

Corporation Information
To find out how many shares a person owns in a corporation, first check to see if that person owns more than 5% of a publicly traded company. Then it should be available in the 10K or 10Q filings that are a matter of public record. If the person owns less then 5% you will have to find out who the trustee and transfer company is and subpeona them. The trustee is also a matter of public record with all corporations.

If you need to find the 10K/Q of a company, one of the best places to do it for free is at: www.10kwizard.com. Just put in the name and you will get all the filings.

To locate the state of a company's incorporation, check the first two digits of their EIN. Chances are that it is the state they were incorporated in.

An alternative avenue is to check KnowX for stocks (knowX.com). It is not free, but it is very reasonable. It will only show ownership of 10% of the company or more, but it will also give indirect ownership type (shares held in a name other than the reporting person but controlled by that person).

To find out when a corporation dissolved and was no longer in existence, contact the Secretary of State's office in the state in which the corporation was charted. Also check: www.internet-prospector.org/secstate.html and www.corpwatch.org. Links to obtain resident agent and registered agent information on corporations can be found at: www.ResidentAgentInfo.com

To find a dissolved corporation in Denver, the information is available on microfiche and public data terminals at the Delaware Department of Corporations at 401 Federal Street in Dover, Delaware. Obtain a copy of the Stock Certificate of Incorporation, Certificate of Dissolution, and Cancellation and any amendments. You can order this by phone using one of their researchers.

In California go to kepler.ss.ca.gov/list.html and enter the name of the company/corporation you need and you will find a lot of different filings for Verizon.

To find contact information for Corporate Resident Agents in all fifty states, as well as Puerto Rico, go to:
www.geocities.com/resident_agent_info/residentagent.html

Company information guide:
www.virtualchase.com/coinfo/index.htm

Other helpful online corporate and business resources include:
- F.E.I.N. search:
  www.freeerisa.com/Customer/login.asp?NotLoggedIn=yes
- Resident Agents by state:
  www.geocities.com/resident_agent_info/residentagent.html
- Forbes People Tracker tracks executives of publicly traded corporations. Just enter name or stock symbol.

www.forbes.com/peopletracker/results.jhtml?startRow=0&t icker=VZ
- Corporate Name Availability: www.lawcommerce.com/research/corp_name_avail_ research.asp
- Business Filings Database: www.llrx.com/columns/roundup19.htm
- Corporation Searches by state: soswy.state.wy.us/sos/corps.htm

A fairly new website with Corporate Reports: www.corporateinformation.com. You will need to register to use this website.

Criminal Record Search

You can search for more than 50 million criminal records from 32 states at: www.rapsheets.com. Credit card is needed for non-member use. Online criminal records are not always current because many counties do not update their records on a regular basis. For a more thorough search you need to hire a local to do a hands-on search if you suspect an employee or business competitor of having a criminal background. When information is required from a less populous, remote county where information is only accessible via a hand search of records, a database service cannot deliver. That is where search companies such as G.A. Public Record Services can assist. www.gaprs.com

The Federal Bureau of Prisons website lists all inmates from 1982 up to the present. www.bop.gov

Master List of U.S. and Canadian prisons with addresses and many phone numbers of facilities can be found at: www.copeministries.org/PrisonList.html

A wonderful source for criminal information can be found on a state-by-state basis on the Wisconsin Department of Justice's website at: www.doj.state.wi.us/dles/cib/sclist.asp#T

Employment Search

A very effective way of locating the current employment of a potential customer or competitor is to run a full credit report and look at the recent inquiries. If you see that a bank, finance company, car dealership, or anyone else that may have pulled his/her bureau to extend credit, send them a "Post Judgment Subpoena for Production of Documents to Non Party," or "Request to Produce Documents" along with a copy of the request going to the debtor's last known address. You can also contact the other creditors and offer to share information with them in exchange for their information.

This is an inexpensive post-judgment discovery tool and most third parties willingly comply. When they send you back the debtors credit application (or documents to support pulling the debtor's bureau), you will usually find much more information than their place of employment.

Suzie McHaney (PI'snoopy@aol.com) of Capital Investigation Services is highly recommended as an expert in locating employment information on hard-to-find subjects.

Dun & Bradstreet has a huge database of public and private companies search-able by principal name, address and telephone number. And it's a free search if you are a D&B subscriber. You can also check to see if the subject is in a licensed occupation through regulation and licensing.

For domain name registration information go to: www.networksolutions.com/cgi-bin/whois/whois and search for company domain names registered to the subject.

For the Dow Jones Interactive Publications Library search Dow Jones' database of 10,000+ news, industry and trade publications for references to your subject. The search returns headlines and a brief abstract at no charge, so the search is basically free. You pay only $2.95 when you obtain the full-text copy of the article.

For Federal Election Commission records go to: www.tray.com and search the FEC's database of political contributors. Many times they provide a scanned image of the form on which the person provides their occupation.

Conduct public record searches, including civil suits, bankruptcy, and UCCs. Many times individuals are named as plaintiffs or defendants along with the company they own or work for. If the person owns a small business, he will have some UCCs against him and the company.

Also check the local Chamber of Commerce. If the person is self-employed, he may be a member of the Chamber. The Better Business Bureau may also have information on the person.

Credit Reports will also aid in locating employment locations. One source for credit reports is www.calcoastcredit.com. It utilizes all three credit-reporting agencies.

Whether you are a lender, apartment manager, pre-employment service, or other verifier, The Work Number makes getting employment and income verifications easy.
www.theworknumber.com

Employee/Employer DoI Tool
Employees and employers now have access to the latest information on government rules and regulations through the World Wide Web. Employment Laws Assistance for Workers and Small Businesses (Elaws) is an interactive Web-based tool that provides expert advisor advice on Labor Department compliance issues, workplace laws, rights and responsibilities, retirement and health benefits, safety standards, and wage and workplace standards. www.dol.gov/elaws

Evidence Eliminator Software
Many companies as well as some government agencies use this Evidence Eliminator software program to protect "trade secrets and sensitive work products." This program is akin to a paper shredder for computers. www.evidence-eliminator.com

Foreign Criminal Records
To locate criminal records for Taiwan/Chinese Fugitives, go to: www.mjib.gov.tw/cgi-bin/crimes/manage-main?mode=general&edition=English

Judgment Collecting
If you have a court ordered judgment, you can legally run credit reports on the person owing, on the basis of that judgment.

In many states, if a judgment debtor fails to pay, an attorney can request that the debtor complete a statement of assets. If the debtor claims he/she has no assets, then the attorney can subpoena the debtor for a post judgment deposition. During this deposition the attorney can question the debtor about income and assets and require that the debtor provide copies of bank statements, tax returns, and other financial records. This process can be expensive. You can also file an Order of Examination in which the judgment debtor has to disclose all assets. Should the judgment debtor not show up for court, the judge will order a bench warrant.

For vehicle related judgments you can file a form with the DMV (DL 30) and have the debtor's license suspended until the amount is paid. If the amount is under $500 then you can file a DL-17 instead. You have to wait until the judgment is final, which is 30 days or 35 days for a default judgment, plus 90 days. You then need to send the original DL form to the court in which the judgment was filed along with a request for a certified copy of the judgment. The court fills out their portion of the form and then mails it back to you. You should then mail the forms to the DMV with their required fee, and they will notify you when the licensee is suspended. Furthermore, you can file another form with the DMV (SR-19c) for proof of financial responsibility. If they have insurance, DMV will send you the information. If they don't have insurance, their license will be suspended. This is mainly for California, but check with your state DMV to see if it applies. Most of the above you can do yourself.

For answers to your questions about judgments go to:
www.bankruptcy-law.freeadvice.com/collections

To find someone who has moved, send mail addressed to them to the old address. In bold letters write ADDRESS CORRECTION REQUESTED, DO NOT FORWARD, or use the postal request form and add "Please check expired files and refer to carrier if necessary." The form and tips on using it can be found at: www.privateinvestigations.org

The website www.autotrackxp.com—called the 4th credit bureau—tracks people who do not use credit cards or who move around.

Ebay is the world's largest auction seller. The person you are searching for may be selling something on Ebay. You can find out by checking: pages.ebay.com/search/items/search-old.html

Verify your subjects address before you attempt process service at this website: www.usps.com/ncsc/lookups

Another method of locating a debtor is to call the local newspaper and find out whom the newspaper carrier is that delivers to the debtor's area. These carriers can be a fountain of information.

Free background check gateway (Accessing Federal Records):
www.backgroundcheckgateway.com/federalinfo.html

Legal Forms and Research

For free downloadable legal forms try *The Electric Law Library* at www.lectlaw.com. It has tons of free legal documents for download. Also try www.courtinfo.ca.gov/forms and www.LawSmart.com

For free and downloadable PDF Legal Forms that you can type into without the need for "form filler" type of software go to: www.strattonpress.com/civpro/civpr07J.pdf

Legal Research by computer can be very expensive if you need to search more than one state at a time. At Versuslaw (www.versuslaw.com) nationwide legal research can be performed for only $6.95 a month or $83.00 a year with unlimited searches. It is better priced than Lexis or WestLaw. You do not need to be an attorney to subscribe.

The Transactional Records Access Clearinghouse (TRAC) at tracfed.syr.edu has put up on the web a breakthrough service for $50 a month that allows users to examine the official actions of individual federal district judges. Using the service accessible on TRAC's subscription site (tracfed.syr.edu/judges), users can review the work of most judges who served from 1986 to 2003. Available information includes judges' workload, sentences, and case disposition times. The service allows the user to compare the work of one judge with the work of all the judges in that district or the nation as a whole and to generate annual case-by-case lists of matters disposed of by a particular judge. Coverage includes criminal cases and civil matters where the government is a party and that were handled by assistant U.S. attorneys.

Medical Doctor's Check
Public records are available in your own state. You can also do a civil litigation search in the county where the doctor practices, identifying him as "defendant." Also check www.questionabledoctors.org and investigative.on.ca/doctor.htm

Military Employment Verification
To verify employment and exact location of a military person call these numbers and they will give you the information. The cost is $3 and it will be sent to you. ARMY: 317-510-2800
USMC: 760-725-5171
USN: 216-522-5301
AIR FORCE: 800-433-0461 #6 or legal 216-522-5301

Standard Form 180 to obtain military records can be found at the National Personal Records Center, Military Personnel Records, 9700 Page Avenue, St. Louis, MO 63132-5100

For other military information, call the Veterans' Administration at 800-827-1000. This number will automatically put your call through to your local VA office.

Passport Info
In order to find out when a person enters or leaves the country, write a Freedom of Information Act request to U.S. Customs for the

Pilot Locating
The Airmen Directory of KnowX.com includes the names of individuals registered to fly by the Federal Aviation Administration. The FAA Airmen Directory requires that individuals must have passed a medical and proficiency exam every two years to be listed. Only pilots who have taken their FAA flight physical in the previous two years, or one year, or six months, depending on class of medical, will be listed in the Airmans Directory.

It is required to have a current medical exam in order to pilot an aircraft, but this does not seem to prevent some aircraft owners from flying their aircraft without a current medical exam. Also try www.loc8fast.com to search for FAA Airmen.

Phone Line Tap Detection and Phone Harrassment
An inexpensive way to see if a phone line is being tapped, is first to make sure the line is clean. Then put on a Time Domain Reflectometer. If you check it every day, you will know if someone is tampering with your line.

Use *57 on your phone when you receive threatening phone calls. This will lock the phone number into the telephone company's computers and most of the time will even lock in the number of one that shows up on your caller ID as being "Unavailable." If you report it to the Police Department they will then contact phone company security, and they will put a trace on that call.

Process Service
Serve-em.com is an internet-based process service-clearing firm serving the legal community with "accountable inter-jurisdictional process service." Up-to-date status is available 24 hours a day, 7 days a week. Upon completion of the service, overnight courier delivers proof of service affidavits compliant with the rules of your original jurisdiction to you. Contact either by going to the website at www.serve-em.com or by calling 800-serve-em or by faxing 888-serve-em. This website also contains free fill in the blank federal forms in PDF format.

Records Resource
Research & Retrieval (800-707-8771) has experienced researchers in virtually every court in the nation. They will not only do court records, but pretty much any kind of document retrieval you need—library archives, any government agency, anything on paper or film. They are very reasonable in price, usually $35 plus copy

costs, and they will do your run, usually the same day, or in at most 2-3 days. They are true professionals.
www.researchandretrieval.com

There is also a company called Research Data (702-733-4990), which operates a pre-pay dial-up service. This service is available to the public.

Search Engine Tip
The Google search engine found at www.google.com allows for different file types to be searched. This will help with locating resumes, power point documents, financial sheets, etc. Use the "Advanced Search" and for your search choose one of the following file types from the "File Format" menu:

- Adobe Portable Document Format (pdf)
- Adobe PostScript (ps)
- Lotus 1-2-3 (wk1, wk2, wk3, wk4, wk5, wki, wks, wku)
- Lotus WordPro (lwp)
- MacWrite (mw)
- Microsoft Excel (xls)
- Microsoft PowerPoint (ppt)
- Microsoft Word (doc)
- Microsoft Works (wks, wps, wdb)
- Microsoft Write (wri)
- Rich Text Format (rtf)
- Text (ans, txt)

Google will also give website information, such as other pages linked to a certain site and web pages that contain the usage of the name. Just type the url into Google for example: www.BusinessSecurity.org and the information will show up.

Security Articles Online
Gary C. Kessler is an Associate Professor and program director of the Computer Networking major at Champlain College in Burlington, Vermont, where he also is the director of the Vermont Information Technology Center security projects. He is also an independent consultant and trainer, specializing in issues related to computer and network security, Internet and TCP/IP protocols and applications, e-commerce, and telecommunications technologies and applications. He has been widely published and has extensive comprehensible articles on security at: www.garykessler.net/library.

Tax Records
Tax Exempt organization numbers are public records and can be verified by calling 202-622-8001. The operator will look the

information up for you. It includes Federal Tax ID Number, asset, and income information.

For U.S. Tax Court information check the Little Black Book at: www.crimetime.com

Another alternative is to go to the Tax Commission to determine whether an individual has filed an income tax return for a particular year. The only information you will see is the name of the taxpayer (or spouse or significant other if filing joint,) and complete address as listed on the return. No other information is available. These records have been known to lead to individuals who otherwise have no "current" address.

Odds and Ends

- Department of Labor - Look at worker's compensation claims processed through the DOL, including the number of claims filed by the individual and details of each claim.
- The National Homeland Security Knowledgebase website covers every organization agency nationwide. Topics include nuclear, biological, bombs, hazardous devices, natural disasters, and chemical emergencies, among others. Its extremely detailed and resourceful website is: www.twotigersonline.com/resources.html
- Monitor your employees with KeyKatcher by capturing the typed key-strokes: www.actcommunications.net/keykatcher
- Most self-storage units of nationwide can be found online at: www.Selfstorage.org
- For research resources for just about anything go to the extensive business section at: www.gate.net/%7Ebarbara/index.html
- For extensive and complete world wide law research go to: www.lawresearch.com/v10/practice/ctindex9.htm
- For offshore banking information go to: www.offshore-manual.com/
- HackerWacker 2000 logs and monitors all user activity on your system, including internet activity. For a demo version: download.com.com/3000-2092-2091645.html?tag=lst-0-3
- For an American Sign Language Browser for your deaf employees go to: commtechlab.msu.edu/sites/aslweb/index.htm
- ReferenceUSA is a database that contains information about ten million U.S. companies. Search by company name, type of business, geographical location, business size, executive name, or ticker symbol. This database includes the most comprehensive White Pages available on the Internet. Great tool, but it has one drawback. You need

Chapter Six — Information Gathering   295

a Salt Lake County Library card number to access the database need to or be using a SLC Library computer. Check it out at: www.slco.lib.ut.us/database-referenceUSA.htm
- The Privacy Foundation website which offers information on privacy-related issues and communications technology with suggested guidelines on how a business can best utilize them is: www.privacyfoundation.org
- For banking information, including offshore go to: www.publicrecordfinder.com/financial.html
- The Americans with Disabilities Act Document Center website provides a wealth of hard-to-find documents on disability law at: janweb.icdi.wvu.edu/kinder
- For AJAX United States and International Government, Military and Intelligence Agency Access go to: www.sagal.com/ajax/ajax.htm
- To acquire a physical address and home phone number of a business contact go to: www.Switchboard.com
- For maps for business owners PDA s go to: www.mapopolis.com/index.jsp
- To recognize the signs of drug usage in your employees go to: www.state.ia.us/government/dps/dne/narcid.htm
- Check out the Swiss Banking Directory at: www.swconsult.ch
- For public record availability laws, electronic court record availability, go to The Reporters Committee for Freedom of the Press at: http://www.rcfp.org/
- For excellent and vast company information resources by the UTPB Library go to: pblib.utpb.edu/company.htm
- Find a newspaper anywhere at: www.newspapers.com
- For scams against businesses check out: www.fraud.org/scamsagainstbusinesses/bizscams.htm
- The website of the Nationwide Court Directory is: www.courts.net
- For copyright & trademark information go to: www.hwg.org/resources/?cid=18
- Census Reports online go to: www.census.gov
- Credit Card Validity Check can be found at: www.qucis.queensu.ca/~bradbury/checkdigit/creditcardcheck.htm
- Internet Usenet FAQ's are archived at: www.faqs.org/faqs
- The Ultimate Travel Portal is at: johnnyjet.com
- For low cost search engine marketing for small businesses go to: www.searchengineguide.com/goetsch/2003/0108_dg1.html
- eBlaster is the keystroke capturing software said to be the best by our computer expert, Kevin Ripa, and is found at: www.spectorsoft.com/products/eBlaster_Windows/entry.asp?affil=1200

Uniform Commerical Code Searches

The Uniform Commercial Code (UCC) is a lien notice filed in the county courthouse and then recorded in a statewide database. The information filed is about loans of almost any type, including furniture, auto, home, etc. The UCC record contains the full name and address of the owner and a description of the property under lien. It also contains the name and address of the lien holder. The lien holder will then have more information on the person who took out the loan with their company, including current location. The UCC records can be searched at the county courthouse where the lien was filed and also statewide by going online to: www.pimall.com/nais/uccr.html

Video Picture Frames

To capture camcorder filmed video pictures on a frame-by-frame basis for evidence gathering against employee or customer theft, check out a software program called VideoWave 5. Experts highly recommend it.

Locating Companies in China
By M. Ettisch-Enchelmaier

I was contacted by an American colleague on behalf of a client to locate two companies on the Chinese mainland and determine their legal representatives. The reason was ultimately to serve legal papers on the companies' legal representatives.

The companies were *"China Power Lighting Ltd."* and *"Chinese Fishing Export Ltd."* [Assumed names for privacy reasons] The names were only supplied in English, with no address, not even a town cited.

First, I tried to develop a telephone number for the companies. Knowing from past experience that knowledge of the Chinese language and a browser needing to be able to read that language were needed, I contacted the international telephone information service. However the Chinese telephone service could not help without any Chinese lettering or place of location.

So I referred the request to long-standing colleagues in both Beijing, as well as in Hong Kong (now part of mainland China). Although they tried to comply with my request, none could help me for the same reasons.

Since I do not take "no" for an answer easily, I decided to search the Internet. I went to the Yahoo search engine and typed "China" in the search box, then "business," and then selecting "only this section." At the same time I opened the free search engine utility Copernic available at www.copernic.com, then clicked on "The Web." I inserted then the name of the first company *China Power Lighting Ltd,* and selected "Search for the exact phrase," and hit the "Search Now" button.

While Copernic was searching about a dozen search engines, I went back to Yahoo, and inserted the name *China Power Lighting Ltd* in the search window. But I inserted the quotation mark (") before and after the name, i.e. "China Power Lighting Ltd." That told Yahoo that no other reading was of interest, i.e. no just *"China"* or *"Power"* or *"Lighting"* or any combinations thereof. Otherwise, I would have been overwhelmed with unnecessary quoted websites. Yahoo did not help in either attempt, but Copernic provided the link I needed to locate the company.

Clicking the appropriate website, I was able not only to get the registered legal seats of both companies, but also information such as their history, boards of management, and lines of business. Furthermore, most important for the American client, the American subsidiaries and legal representatives were also listed. This meant that I could most likely serve the companies via their representatives in the United States, thus saving time and money. With the transmission of this information I was able to successfully close the case.

Conclusion: This was a case where the traditional sources such as databases, commercial register, and telephone directories were unable to help. Only the new medium—the Internet.

Corporate Intelligence Collection Process

By Ken Wold

One of the most important facets of intelligence is to reduce the ambiguity inherent in observing external activities. What is intelligence? It is collecting, collating, evaluating, analyzing, integrating, and interpreting. It is a product which will give the collector specialized information to further his aims or goals. In the case of a country, the most obvious use of intelligence is its use by an adversary to gather information concerning military weaknesses, political direction, and economic programs. In the case of a company, the intelligence foe may be a direct competitor, a new market emergence business seeking an "edge," or current employees attempting to strike out on their own.

The intelligence sought may provide that persons or entities with just the information needed to develop their strategy to reach their goals and at the same time to impair or obstruct yours. As with any intelligence, you must collect it from a number of sources to gain the best data and information possible to make informed accurate decisions on the course to take. Initially, the data received is very raw and unfinished. It must be processed, analyzed, compared, tested, and conclusions drawn before it can be utilized for its intended purpose. Just as you would follow a strategic process to form a division, create a new product, or make management decisions, you must conduct analytical review of the intelligence data collected. If all sources of intelligence are utilized, the risk of errors or deception is dramatically reduced. A brief overview of the intelligence process is needed to understand how it is accomplished and structured.

There are several means by which intelligence is collected and disseminated—humans (witting or unwitting), signals or communication, images, and measurements. In the corporate world intelligence probes and violations are an everyday occurrence. In the last few years various attempts that have been made to prevent, detect, and neutralize intelligence threats have proven to be somewhat successful. As technology develops, the hunger and economic necessity to stay ahead of the competition also increases. Corporate intelligence security and training on how to recognize threats have improved somewhat. However, most employees continue to cause poor physical, operations, and communication security. Competition and adversaries use each of the above disciplines to collect information. In the U.S., which is an "open" society, intelligence activities are very lucrative to operate. Online databases have underscored this weakness and provide tailor-made information to those willing to seek it out.

Operations security is paramount to insure that protective countermeasures are taken to prevent compromises. The best and easiest manner in which to collect intelligence information is still the old-fashioned way—human intelligence collection. Other means utilizing technology, such as optical, audio or a combination thereof, works well in tandem with the human collector. Technology, however, has a cancellation effect. A new hybrid emerges and another is created to cancel it out. One of the keys to success in using humans is access. In most cases a collector must have access to the area where the information or data is stored or maintained. The most technologically advanced image and eavesdropping equipment are totally useless unless you have access. Once access is gained, the tandem of human and technological intelligence is virtually unstoppable.

Many people tend to believe human intelligence is associated with only espionage. This is incorrect. An example would be the conversation people have with each other on a daily basis—questions such as, "What are you working on?" or "Hey, did you know the company was planning to move up production of that new engineering process?" These are the kinds of questions people hear each day in their work routines, never giving a thought that the co-worker could be an intelligence collector for a competitor. Why are they not aware? Because most are not trained to be alert for such situations or carry a laissez-faire attitude toward physical, operations, and communications security. There are many types of human collectors.

Human forms of intelligence collection include overt, clandestine, and exploitative. Overt intelligence collection is conducted openly and is of the type(s) discussed earlier—office discussions, social engagements, and willing participants. Some of the overt collections can occur by attending meetings, conferences, volunteering for projects, seeking lateral company movement for a certain position, joining associations and organizations, increasing social circles, and garnering awards. Disclosure of their identity will not compromise the collector or the sponsor.

Clandestine sources, however, seek to remain anonymous at all times. These types of human collectors have been recruited to give information to their sponsor. Identification of the source or the sponsor would prove to be a terrible embarrassment. There are no legitimate estimates on how many clandestine operations are occurring against a typical corporate entity. However, I can state with certainty that clandestine collections *are* taking place and at a rate higher than a typical corporation believes. Much of this clandestine work emanates on a global scale and not simply in the U.S alone. The espionage activities of many countries concerning military intelligence only has been replaced by technical and scientific exploitation. Due to the openness of the United States, it has more conferences, science and technical officials, and open records than any country in the world. Obviously, it is one of the premier targets for corporate espionage collection. Any clandestine collection operation requires an extensive support base with many assets. As with any operation, the more personnel involved opens the greater opportunity for errors to be made and detection to be somewhat easier.

Exploitative intelligence collection is a means to obtain information and data when other means have failed. Exploitative collections are similar to clandestine operations since they are covert in nature and rely on the cloaking of identities and purpose(s). However, individuals and entities tend to exploit people when other attempts to recruit them seem to have failed. This method involves offering bribes, compromising positions, using threats, and force, kidnappings, extortion, and many other attempts to gain information. It is the last resort for the sponsor of such tactics, but it can be very effective. Remember that an intelligence adversary will resort to the most appropriate tactics in order to attain their desired goal. Many of the human intelligence collectors or sources will utilize signal intercepts to assist them in obtaining data or the information they need.

Signals intelligence incorporate interceptions of communications and electronics (computers). Communications include any voice emanation equipment. Electronics involve emitting transmissions from non-communications sources. Intelligence collection from signal sources can be accomplished in many ways—on land or on water, fixed or moving, or via any electro-optical/audio device capable to intercept that particular signal. There are simply too many interception devices available to individually identify here. Imaging is another excellent mode of intelligence collection.

Imaging intelligence is accomplished by electronic or optical means on film, displays or other electronic media. With the advent of digital imagery, intelligence collection takes on a new form. An electronic digital source can be manipulated with little or no degradation in imagery. A whole host of new deceptive techniques by an intelligence sponsor includes digital photography and motion picture manipulation to attain the necessary means of embarrassing a company, as well as extortion and creating illusory, suggestive techniques for their own company or product. In other words, placing someone in a compromising position by means of a deceptive picture or video. The digital age has given imaging collectors an exciting and intriguing future. Space-based imagery systems are proliferating quickly. They provide a platform in which every area of the earth can be viewed and recorded with the precision of viewing microbes through a microscope.

Governments are not the only entities who operate space-based imagery systems. Commercial systems are now operating in the space environment with much of the same technology as government equipment. However, the commercial applications, as the name implies, results in images being distributed to those willing to pay the price requested. This is a very dangerous precedent. There are some limitations to imaging systems. Imaging is normally degraded in darkness and adverse weather with the exception of night vision and thermal imaging devices. Electronic countermeasures can be instituted to degrade or defeat imaging equipment by providing misleading images, scrambling, camouflage, concealment and other means. Effective deception techniques employed will make sure the collector receives erroneous data and information.

In any process you must first plan what you want to do. Direction is needed from identification of what intelligence is required to the end result of the finished product. Whenever an individual or entity determines a need for intelligence, they will identify, prioritize, prepare, collect, disseminate, and survey the situation constantly. The intelligence seeker must collect, determine weaknesses in the target, and select from among its assets for the collection process, as well as their time to collect the information or data sought.

Collection programs are initiated and requirements developed to meet the sponsor need. Based on these requirements, specific tasks are needed to accomplish collection. Collection is done by a myriad of collection techniques and disciplines. Success and failure are both built into the collection scenario, so alternative plans need to be in place. Assessments are made using different types of information, sources, and methods. Any collection effort must be secure, rapid, reliable, and complete. Information collected must be compiled and forwarded for processing action.

The information gleaned must then be processed so someone can analyze it all. It may need translation, conversion from text to imaging/audio, or vice versa. Video production, correlating data, and processing it for final production is required. Producing the information collected involves evaluating it all and providing a finished product to the sponsor. The final destination of this entire process is to disseminate what has been collected to the sponsor. It is normally accomplished in a secure manner, such as written form. The sponsor may determine the data collected isn't sufficient and will require further intelligence gathering efforts.

Any corporate security and new employee program should incorporate an awareness program for all personnel and provided a refresher training on a periodic basis. Having a solid foundation of the intelligence collection process can assist in preventing the opposing collector from obtaining information. It provides corporate security managers with ideas for protective measures, whether their facility may or may not be a target by a certain individual or entity. Having knowledge of collection methods and efforts will also serve to establish a program to dissuade intelligence collection efforts and build deceptive techniques.

Ken Wold is the Founder and President of Guardian Investigations, a private investigations firm based in Boise, Idaho and the Investigator Network (I.N.), a worldwide investigative association providing data/resources/information. He is also the author of "The Complete Investigator's Training" available on CD-ROM for $29.95 or free with membership at I.N.

Finding a Reputable Private Investigator

By T. A. Brown

In today's climate of unease, people are affected everywhere, especially in the business community. Business owners need to be especially careful of the people they hire as well as the people they do business with, including vendors and customers. With that in mind, how does a business owner find and hire an ethical Private Investigator (PI) who will do a good job and who is reputable?

A very high percentage of licensed PI's are former or retired Law Enforcement Officers, FBI, Secret Service, or Military personnel, and with the exception of the few states listed below who don't require licensing, have to go through extensive training in order to qualify for licensing in their state. Looking in the yellow pages for a private investigator is kind of like being a kid again and saying eenie, meenie, minee, moe. Additionally, bigger ads do not mean better PI's. Many good PI's maintain a low-profile image.

These are the states that do not require licensing for a private investigator—Alabama, Alaska, Colorado, Idaho, Kentucky, Mississippi, Missouri, South Dakota, and Wyoming. In-house company investigators who do not work outside of the company may not be subject to licensing laws in the remaining states. Check with your state for in-house investigator requirements if you are considering implementing such a department within your company.

Here are the most important opinions of nearly 2,000 PI's that I polled on how they would go about finding a reputable private investigator, and as you can see, these professionals hold themselves up to the highest of standards:

- Ask your local highest-ranking law enforcement officials. They have very likely run into PI's, both good and bad, in the course of their job and may have opinions on who they would deem to have quality and ethics, and who to stay away from.
- Ask an attorney, paralegal, or even the legal secretary, who may just know better than her boss.
- If you know of other people who use PI's frequently, ask their opinion. Word of mouth is often the best advertising for a good investigator.
- Interview the PI's you are considering using once you get

some referrals. Ask these types of questions: How long they have been licensed? What type of cases do they specialize in? How much training have they had in the type of case you need them for (don t hire a traffic accident expert for a criminal defense case)? If they don't work your type of case, will they recommend someone who does? Are they qualified to testify in court in cases like yours if necessary? If you are working with an attorney, have they worked with your attorney or the opposing attorney before?

- See if they appear professional in their dress and manner, not slovenly and disorganized.
- Make sure you feel comfortable and compatible with the PI you select.
- Always ask their fees up front and have them list what their expenses are that you are required to pay for and how often they require those payments. If they are evasive in telling you, remember "caveat emptor." An honest PI will tell you right up front what their fees are and what their time is worth. Cheap is not necessarily good and expensive is not necessarily better. One PI explained the charges for their services very well when he said, "People will pay a plumber $75 an hour to change a water heater, but complain about paying an investigator $100 an hour to risk his life hunting down witnesses or bad guys."
- Ask for references and then follow through by checking them.
- Never hire a PI who guarantees results, but rather one who offers a contract that states what they will do for their fees. In that contract should also be a statement that you will receive detailed reports for the time spent. You want to see what you are paying for.
- A description of the case should also be included in the contract, and you and your investigator need to discuss ahead of time how to approach your case. If there are special circumstances that will dictate this approach, you will need to make your investigator aware of it. Observe if the investigator is seriously listening to your advice in handling a sensitive issue.
- Please remember that the investigator you are hiring is not in the handholding business, and he will have many other cases he is working on besides yours. The time you spend on the phone with him is costing you money and him time, much the same as talking to an attorney. A busy and reputable investigator will definitely charge you for all phone calls after the initial free consultation. This is a business expense and will come out of your retainer. On the other hand, there may be times when it is imperative that

you be able to reach your investigator. Will he/she provide 24/7 accesses?

- Find out if the PI is a member of the local Chamber of Commerce and what other organizations he or she belongs to. Check the Better Business Bureau to see if there have been any complaints filed against him or her, or anyone else in the agency they own or where they are employed.
- Check the courthouse records to see if disgruntled clients have sued the PI or anybody he or she works with.
- Contact the State PI Association in the state where they work. The Secretary of State will be able to give you that contact information or you can find it listed on the Internet at www.pimall.com/nais/pi.assoc.list.html. The State Licensing agency will also be able to tell you if any complaints have been filed against the PI's agency or the individual PI in question.
- Find a PI who is computer savvy and has an email address as well as a fax machine. Computers play a very important role in today's investigations. The days of Sam Spade are no longer enough, although legwork is still a necessary part of it.
- Make sure that the PI is fully covered with verifiable Errors & Omissions insurance to avoid any possibility of your getting in the middle of a lawsuit if your investigator makes a serious sueable mistake.
- Be wary if an investigator assures you that he can guarantee the outcome of the case. Things often are not as simple as they seem on the surface and can change at a moment's notice. A wise investigator knows that.
- Make sure that the contract has a confidentiality clause and that the party being investigated will not be informed of the investigation. The exception to this would be a signed release of information form from an employee or possible new employee being considered for a job. Most PI's will readily guarantee confidentiality. Be aware, however, that the confidentiality between yourself and the investigator can only be guaranteed as far as the law allows. A PI can be subpoenaed to testify in a court of law. Check your state laws or ask your attorney. It is your responsibility to find out and know the law and be prepared.

Finally, there are a few professional organizations that can help you in your quest to find the most qualified investigator anywhere in the world for your needs. NCISS, is an organization that all PI's are eligible to join and is their lobbying venue.

National Council Of Investigation & Security Services
NCISS Headquarters Washington D.C.
1730 M Street, N.W., Suite 200
Washington, D.C. 20036
www.nciss.com/

National Association Of Investigative Specialists (NAIS) World's largest association of profession private investigators with members from around the world. Over 3,500 plus members where you can search for them by state, city, country and look at their various investigative areas of expertise. www.pinais.com

These other two organizations are by invitation only, and each PI is thoroughly checked out before he is considered for membership. These would be excellent sources for highly qualified private investigators.

W.A.D
World Association Of Detectives
Brough, HU15 1XL, England
Ph: +44-1482-665577
Fax: +44 870-831-0957
wad@wad.net
www.wad.net/

INTELNET
International Intelligence Network
P.O. Box 350
Gladwyne, PA 19035
Ph: 610-687-2999 or 800-784-2020
Fax: 800-784-2020
intelnet@bellatlantic.net
www.intelnetwork.org/about.asp

The Right to Privacy

By Mike Hawkins

In today's increasingly more technological society more and more people are becoming worried about their loss of individual privacy and fear that George Orwell may have been wrong about the grim future when he wrote his novel *1984*. What rights and protections do we as citizens have against unwanted intrusions into our lives? In fact, do we have any such rights at all? Or are our lives to be lived as an open book ready to be read by anyone who wants to look?

Data is collected on us every time we make a credit card purchase, or file insurance, job, and other applications. The phone companies (wire and wireless) know our calling habits and even who we call. However, most people accept these forms of information gathering without concern, but they are concerned about their privacy at home and in their vehicles. The topic of "privacy" is far too detailed and broad to cover in a single article, so we will limit this discussion to the general concepts of privacy and your right to, or your loss of, privacy while moving in public.

It must be understood that in the United States we live in a society comprised of fifty-one sovereign governments. Each of the 50 states is a government with law-making powers over its citizens and everyone within its borders. In addition, we have the federal government which has similar powers over everyone within its national boundaries. The difference is that the federal government can establish limitations on "governmental" intrusions into our lives. Each of the states can further limit government intrusion, but cannot violate or reduce the federal standard. Since each state is sovereign and different, this article will discuss only the federal standards which all states must adhere to.

Privacy issues are considered in a criminal law context rather than a civil law one. That is why in the Bill of Rights the Fourth Amendment sets limits on the powers of the government, but not on you or your neighbors. The Fourth Amendment states: "The right of the people to be secure in their persons, houses, papers and effects against unreasonable searches and seizures shall not be violated and no warrants shall issue but upon probable cause supported by oath or affirmation, particularly describing the place to be searched and the person or things to be seized."

This amendment limits the power and authority of the government to conduct only *reasonable* searches and seizures which must be based upon a warrant supported by both probable cause and sworn to under oath or affirmation. Over the years there have been many judicial rulings interpreting the meaning of the amendment and creating some exceptions to the warrant requirement. For example, when exigent circumstances exist, the warrant requirement can be overlooked but not the need for probable cause which will still have to be judicially determined at a later time. One of the most common examples of such an exigent circumstance is that dealing with vehicles. The courts have ruled that since a vehicle is mobile, it does not stay in the same location and may be impossible for law enforcement to relocate. Therefore with probable cause, it may be searched without the necessity of obtaining a warrant.

This general concept must be viewed in light of a number of cases which have dealt with what has become known as the *reasonable expectation of privacy* (REP). The courts have ruled that we may have REP, even in public, if certain key standards are met. The test for REP is rather simple and addresses two questions:

1. Has the person demonstrated his or her intent to keep something private?
2. Is society willing to grant that privacy?

Both questions must be answered with a "yes" for REP to exist. By way of example, in the leading case of <u>Katz vs the United States</u>, Mr. Katz conducted illegal activities utilizing a pay phone which was inside an old fashioned phone booth. Federal agents unable to obtain a warrant for a wiretap placed a hidden microphone in the booth so that they could capture and overhear Katz' conversation. The court ruled that by entering the phone booth and shutting the door, Katz demonstrated his intent to keep the conversation private and that privacy was something that society was willing to grant him. Therefore, the contents of the conversation could not be used against him even though Katz was in a public place.

By enacting certain laws Congress has broadened the concept of privacy beyond that which would be required under the Fourth Amendment. Examples of such legislative-created privacy protections include the statutes prohibiting the interception of cordless and cellular telephone communications. Under a normal Fourth Amendment/Katz analysis of such phone use, once the signal is released into the air via radio waves one would lose any REP and such communications could legally be intercepted. This is because society is not willing to grant privacy to openly transmitted radio signals.

As we go about our daily activities, what privacy interests do we have to protect us from unwanted intrusion and spying? In most instances we can rely upon the REP standard, which is a basic minimum amount of protection. This means that once we leave home and either walk or drive down the street we have no REP that we will not be observed or followed. It is this lack of REP that allows the police or private investigators to "tail" people. Of course, this right to follow anyone in public has recently been curtailed and limited by the numerous "stalking" statutes that have been enacted by state governments.

In reading comments and questions placed on the internet on different lists and forums for private investigators, it is apparent that the ability to utilize electronic tracking devices placed on vehicles is a concern both to the investigator and the person being followed. As a general rule of thumb (again based on the federal standard), such devices may be placed on a vehicle and used to follow it in public. However, note that when the device is placed on a vehicle, the vehicle must be in a public place or permission must have been obtained to enter the location to place it. One may not enter a subject's garage, for example, to place the device since the garage itself is protected from such intrusions. Likewise, the device may only be placed on the exterior of the vehicle as REP exists in the interior of the vehicle. This concept also applies to the actual "tailing" of the vehicle which can be done only while the vehicle is moving about in public on highways and streets. Once the vehice enters an area where REP exists, it may no longer be followed either in person or electronically. Remember that there are other possible problems with utilizing such devices as the result of various state law and court decisions. These will create problems for the investigator while giving greater privacy to the person being followed.

Although we have REP in our vehicles, the degree of that REP is reduced due to a number of factors such as when vehicles move on public lands, are licensed, and are more readily observed by other members of the public. Therefor, there is generally no REP with regard to the exterior of a vehicle, although we do retain a reduced REP with regard to the interior. The fact that vehicles are in public and can usually be easily looked into further reduces the degree of REP. In a normal car, for example, there would be no REP that something sitting on the back seat would not be observed, while there would be REP for items kept in the glove compartment or trunk.

This concept of REP is applicable everywhere. When we enter a public restroom, we expect a higher degree of REP in the stall than we do while standing at a row of sinks. When we use a pay phone attached to the wall as part of a bank of such phones, we have no REP that our voice will not be overheard. When we stand in our front yard, we have to expect that our actions can be observed, while when we are inside our homes we can expect a much greater degree of privacy. It should be noted however, that if I am in my home with my drapes and blinds open, I must expect that someone on the public sidewalk will be able to look into my house and observe my activities. However, should someone walk up to my window to peer through a crack in the blinds, my REP has been violated.

Under the federal standard, if someone in an apartment across the street can see into my home with the naked eye and observe something, it would be both legal and admissible as evidence. In fact, if something can be observed with the naked eye, it can also be enhanced by the use of a flashlight, binoculars, or other such device. However, if it could not be observed with the naked eye, it may not be enhanced. For example, someone a long distance from my home may be able to see my windows but not see into my house. That person would be prohibited from utilizing a powerful telescope or other device to look into my home since it was not initially visible by the naked eye.

There are many other privacy issues that face us today. Another one that is of great interest is the listening in on or the recording of telephonic communications. The federal standard is that if one party to a communication consents to its being recorded or intercepted by a non-party, it is legal. A number of states, however, have passed legislation requiring that consent must be obtained from all parties for such recording or non-participant interception. This means that I cannot tape or tap my own phone while my wife or children are using it without their consent. Violation of this law carries some very stiff and penalties, not only for the person doing the tapping, but also for anyone who attempts to use any information gained by such interception.

While we each have REP in our personal property and belongings, that REP can be abandoned by our actions. Abandonment in a REP context does not follow or track with the normal property law concepts of abandonment, in that one does not have to abandon the property, just one's REP in the property. For example, when a police officer asks a person what is in the bag he is carrying and the response is "I don't know, it is not mine," that statement indicates that he has abandoned his REP in the bag and it can be searched without a warrant or probable cause. It should be noted, however, that REP is a *personal* right and generally can not be waived by anyone else. This means that if the person was merely carrying a bag for someone else, any evidence or contraband found inside *might not* be admissible against the actual owner of the bag.

One way people find out about others is going through their trash and garbage. The U.S. Supreme Court has ruled that once the garbage is placed on the curb, all REP has been abandoned since one cannot expect that homeless people or the police will not go through it. However, if the trash can is kept in the back yard, the can itself is within an area of REP and cannot be gone through or searched. A few states have provided an even greater degree of privacy interest in such trash than has federal law, so before you "dumpster dive" it would be wise to check your state laws on this matter.

A similar situation and one with even greater potential for problems for individuals is the abandonment that occurs when we throw away our credit card receipts in the store trash can. The collection of a credit card number under such a situation is perfectly legal, but if it is used to purchase other items or services a crime has been committed. However, it is one way for criminals to get information necessary to steal one's identity, and in most cases identity theft is not considered a crime. This is true even though the theft of one's identity can cause greater problems than the illegal use of the card to make unauthorized purchases.

When it comes to jointly owned or used property, the general rule is that either party may give permission for a search, even over the objections of the other party. A typical situation is that either a husband or wife may consent to a search of the common areas and commonly used items in a house, but may not consent to an area or item that is strictly under the control or use by the other party. An example of this would be a husband who maintains a home office with filing cabinets, etc. that only he uses. His wife cannot consent to a search of the filing cabinet since it is under the exclusive use and control of the husband. The law in this area gets somewhat trickier and more confusing when we are talking about parents and children. Generally, a parent may consent to the search of a child's room since the parent provides that space and has the right to enter it at any time. However, if the child is older and pays rent, then the parent has lost that common usage and may not consent to a search.

There are many other factors that may come into play and each case must be determined on its own unique set of facts. For instance in a town outside Atlanta, police went to a teenage suspect's home and asked his mother for permission to search her son's room. Her reply was, "I guess you can, if we can get in." the police asked her what she meant, she said "My son is a very private person and has padlocked his bedroom door, so his father and I are not allowed inside." In this situation one could argue that the mother had no right to consent to the search since it was under the exclusive control of her teenage son. On the other hand, it is possible that since the room was part of the parent's home and they had given him such privacy without the payment of rent, they could also revoke their decision to permit such privacy and recoup the commonality of the son's room. Should a parent be able to revoke the benefit of privacy, this would then make the mother's consent valid.

One should always keep in mind that the Fourth Amendment limitations and its progeny are applicable to *governmental* action and do not limit the actions of private individuals. This means that should a burglar break into a person's home and find evidence of criminal wrongdoing, such as possession of child pornography, the burglar may turn that evidence over to the authorities and it will be admissible. This is so because the intrusion into the home was done without the knowledge, consent, or cooperation of the government. This means that despite one's constitutional protections against governmental intrusion, the only prohibition against *private* intrusions are based on both criminal and civil statutory and case law. In the previous example, the burglar is guilty of the felony crime of burglary and may be prosecuted for that crime, but the evidence he found will be admissible against the homeowner. Hence, your privacy is, in effect, based on the care that you take to keep things private.

A former South Carolina State Constable, Mike Hawkins was admitted to the South Carolina Bar in 1972. He practiced law for fifteen years as a trial lawyer in both State and Federal Courts. After closing his practice, Mike became an Agent for the U.S. Naval Investigative Service and conducted felony level criminal, espionage and counter-terrorism investigations in the U.S., Europe and Central America. He then became a Supervisory Criminal Investigator at the Federal Law Enforcement Training Center where he trained thousands of Federal, State, Local and International criminal investigations students. While there, he was selected by the U.S. Departments of Justice, Treasury and State to be an instructor in Law Enforcement Management and Leadership at the International Law Enforcement Academy in Budapest, Hungry. After leaving Federal service, he became a Senior Judicial Educator for the State of Washington before becoming the Chief Legal Counsel for a government contractor responsible for the detection and investigation of Medicare fraud in thirteen states and three Pacific territories.

Dumpster Diving, ie: Garbology
By T. A. Brown

Why would a business owner need to know about Dumpster Diving? One reason to dumpster dive could be to discover information about your competitors. If they are not smart enough to shred their documents (which is advisable for all business owners), a lot of confidential information can be learned about a company just by sifting through their trash.

Another reason to dumpster dive would be to locate bank account records, assets, and spending habits from a deadbeat debtor. Bank account and credit card records are difficult to obtain because of privacy laws, but the garbage bins are fair game for discovery purposes. Joseph Culligan, author and recognized public records expert, did a television exposé on a well-known bank's current practice of discarding records without shredding them first. Your very own personal bank records of recent deposits, withdrawals, and other activity could easily be sitting in the dumpster behind your bank as you read this. This means that if you search the personal trash of the person you want to collect from and in the process discover the name of the bank the person does business with, as well as his bank account number, you might get lucky and find his bank records in the bank's dumpster. This information could help in collecting the debt by attaching their accounts.

When is it legal to pull someone's trash and is it legal when that trash is placed in a dumpster?

Professional "garbologists" point out that although the law varies with different states, it is legal to obtain someone's trash once they discard it and it is placed outside their property for pick up. However, you have to keep in mind the location of the dumpster since you might be trespassing. The key to the trespassing issue on dumpsters is public access. Make sure that the dumpster is not on private property, rented or otherwise. The dumpster does not have to be in a gated area to be on private property. Public access must be available.

Once property is discarded and put on the curb and is no longer on private property, it is considered abandoned property and is free for the taking and acceptable as evidence in any court in the United States. Attornies have even requested their investigators do trash pulls on the opposition.

All "evidence" found needs to be bagged, labeled, and locked up to maintain the chain of evidence for court. When going through trash, wear heavy duty gloves for your own protection because you might come in contact with needles and syringes. Also, bring replacement trash bags the same color of those you will be searching so that you can replace them and search them at another location without the subject suspecting.

Employers should be especially careful about shredding unwanted job applications that have been submitted to them. Social security numbers and a wealth of information can be found on these documents that have been entrusted to you.

Trash is one of the best and most legal means of gathering information. Even the government and law enforcement use it. Old credit card bills, utility bills, phone bills, etc. are all possible finds. Lots of useful information can be found in abandoned property searches. As always, please check your applicable state laws as they vary from state-to-state.

The legal decision on "dumpster diving" can be found online at: caselaw.findlaw.com/scripts/getcase.pl?court=US&vol=486&invol=35

For more legal information about the topic go to: www.rbs2.com/privacy.htm#anchor666666.

ACKNOWLEDGMENTS

First and foremost, my very special thanks go to Joseph Culligan, who has accepted my family and I with his kind patience and friendship. A renowned author as well as a nationally known expert private investigator, he has sent my life down paths I never dreamed I'd walk, and this book would absolutely not exist in deed or in thought without his unintentional and unassuming quiet, positive influence. He is a good man and a gracious gentleman. Thank you, Joe.

I especially want to thank all the private investigators, former law enforcement officers, retired military and FBI, professors, attorneys, and other security minded individuals who have contributed articles for this book. They are great writers and I appreciate their willingness to help teach business owners by sharing their expertise.

The biggest thanks goes to Carl Heintz who encouraged me down the path that this book has become. He took the time to read a raw manuscript from an unknown author, saw a potential business book, and helped mold it with an entirely different focus than it started out having.

I would be remiss if I didn't acknowledge the "wonder man" editor of this book, Dr. Charles Patterson. What I thought was a perfect book became an edited piece of art. He's very good at what he does and his sense of humor made it an easy and enjoyable process, and his handwriting was even legible. His website can be found at www.excellenteditor.com.

Many thanks go to Kathy Legg of www.LittleBrownMouse.com for the time she spent in helping me learn to fix software problems. Thanks to my son Dan Brown and to Chris Davis for their help with website technical setups that were beyond my comprehension. Special thanks to my good friend Nancy Wagley of Pennsylvania for her diligence in the tedious job of verifying the activity of many of the links included in this book. Appreciation also goes to my many cousins for their fun help with flyer mailings. And last, but far from least, for her unwavering support of me in all my small ventures and her willingness to help with the marketing this book, thanks go to my daughter Diane Brown. She's one smart cookie.

www.BusinessSecurity.com

Author Contact Information

Brown, T. A.
Former editor / publisher of the Joseph Culligan newsletter.
Freelance writer & researcher
tabrown@BusinessSecurity.org

Benavidez, Treyce
The Handwriting Company
P.O. Box 151468
Austin, TX 78715-1468
512-481-2777 Fax: 775-924-3071
HandwritingCompany@yahoo.com
www.Treyce.com

Castleman, Scott R.
www.ccca.homestead.com
crimecontrol@juno.com
503-920-1906

Collier, Brenda H., Esq.
Glast, Phillips & Murray
Galleria Tower I
13355 Noel Road, Suite 2200
Dallas, Texas 75240
972-419-8316 Fax: 972-419-8329

Cook, Gary R.
Security Design Sciences
9628 Coosa St.
Ventura, CA 93004
805-659-1952 Fax: 805-659-4123
Cell: 805-208-2938
www.homestead.com
  /securitydesignsciences
Gcooksds@aol.com

Culligan, Joseph
Miami, FL
www.JosephCulligan.com

deKieffer, Don
deKieffer & Horgan
729 15th St., NW
Washington, DC 20005
202-783-6900
www.dhlaw.com
ddekieffer@dhlaw.com

Earnshaw, Joan
PO Box 573
Flora Vista, NM 87415
505-334-2206
joanearnshaw@cyberport.com

Elliott, John M.
Elliott Consulting Group
Calabasas, CA
818-889-4771
elliott@elliottconsultinggroup.net
Member: Society Of Former FBI
  Special Agents

Ettisch-Enchelmaier, M., B.A.
Internationale Wirtschaftsauskunftei,
  Detektei
Bodelschwinghstr. 9
67246 Dirmstein/Germany
Ph: (+49 6238) 989 098
Fax: (+49 6238) 989 099
Ettisch-EnchelmaierGmbH@
  t-online.de

Gardner, Robert
Security, Crime Prevention &
  Community Safety Advisor
PO Box 6880
Ventura, CA 93006
805-659-4294 and
2620 Regatta Dr., Suite 102
Las Vegas, NV 89128
702-733-8711 Fax: 805-659-4159
www.crimewise.com
cpp@crimewise.com

Goldmann, Peter
213 Ramapoo Rd
Ridgefield CT 06877
800-440-2261 Fax: 203-431-6054
www.wccfighter.com
editor@wccfighter.com
Editor of the *White-Collar Crime
  Fighter*

Hawkins, Mike
The Hawkins Group
22410 95th Place West
Edmonds, WA 98020
425-775-3395
mikehawkins@hawkinspi.com
www.hawkinspi.com

Keltner, Brian
L. A. Confidential Investigations
539 E. Bixby Rd., #35
Long Beach, CA 90807
877-689-3687
bkeltner@l-a-confidential.com
www.l-a-confidential.com

Klein, Barbara
The Profile Agency
Pager: 877-390-4897
4800 Whitesburg Dr. #30-225
256-533-0388   Fax: 256-533-3259
Huntsville, AL 35802
ProfileA@aol.com

Kuebler, Col.(Ret) David W.
Blue Knights Investigative Services
P.O. 640877
Kenner, LA 70064
504-416-7400   Cell: 504-416-7400
dwkuebler@cs.com

McCourt, Michael G.
508-833-7171
mgmassoc@adelphia.net

Pankau, Edmund J.
PO Box 9797
The Woodlands, TX 77387
713-224-3777 or 800-352-6519
Fax: 713-236-0494
ejp@pankau.com

Pruitt, Alan CPP
PO Box 5328
Pleasanton CA 94566
877-379-9261
www.alanpruittcpp.com

Rahn, Charles T.
A Very Private Eye, Inc.
Orlando, FL
407-273-6646

Repa, Barbara Kate, Esq.
1940 Scott Street
San Francisco, CA 94115
415-441-4203   Fax: 415-441-6070
BKRepa@yahoo.com

Riddle, Kelly
Kelmar and Associates
2553 Jackson Keller, Suite 200
San Antonio, Texas 78230
210-342-0509   Fax: 210-342-0731
kelmar@stic.net
www.kelmarpi.com, www.a-c-i.org

Ripa, Kevin J.
J.S. Kramer & Associates, Inc.
Computer Evidence Recovery, Inc.
42219-400 9737 Macleod Tr. S.
Calgary, Alberta, Canada T2J 7A6
403-861-4846   Fax: 403-271-0186
Toll Free: 877-861-4846
www.jskramerpi.com
info@computerevidencerecovery.com

Roberts, David P., FIPI.
British American Consultants, Inc.
200 Gregory Place
West Orange, NJ 07052
973-324-2395   Fax: 973-324-2396
basec007@aol.com
www.britishamerican.net

Russell, Jeffrey A.
www.ncisinc.com
jar@ncisinc.com

Schmedlen, Roger H.
Loss Prevention Concepts, Ltd.
35560 Grand River Ave., PMB # 311
Farmington Hills, MI 48335
810-632-6636
expert@lpconline.com
www.securityexpertonline.com
www.lpconline.com

Sheffer, Todd
Trade Intelligence, LC
P.O. Box 895
Mt. Airy, MD 21771
301-829-9384
tilc@adelphia.net

Siciliano, Robert
POB 15145
Boston MA 02215
800-2-GET-SAFE   Fax: 877-2-FAX-NOW
www.StreetSafeSecurity.com
Robert@SafeTravelSecurity.com

Smiley, Christina
Sierra Judgment Recovery
PO Box 9258
South Lake Tahoe, CA 96158
530-542-2102   Fax: 530-542-3084
CSmileySJR@prodigy.net
www.recoverycourse.com

Spear, Robert K.
410 Delaware
Leavenworth, KS 66048
913-682-6518
www.sharpspear.com
sharpspear@kc.rr.com
Sharp Spear Enterprises and
   The Book Barn

Spence, Justin
407-447-0085   Cell: 407-928-9504
LnPDetective@msn.com
www.SigTruth.com

Widell, Coleen and Naylor, Rick
American Institute on Domestic
  Violence
2116 Rover Drive
Lake Havasu City, Arizona 86403
928-453-9015
www.aidv-usa.com
info@aidv-usa.com

Witherspoon, Ralph
Witherspoon Security Consulting
22021 Brookpark Rd. #100
Fairview Park, OH 44126-3100
440-779-3803   Fax: 440-779-3203
witherspoon@security-expert.org
www.security-expert.org

Wold, Ken
Nampa, ID

Zalma, Barry, Esq., CFE
Law offices of Barry Zalma, Inc.
4441 Sepulveda Boulevard
Culver City, CA 90230-4844
310-390-4455 Fax: 310-391-5614
Bzalma@earthlink.net
www.zalma.com

INDEX

AAA, 22, 105
Aachen, 92
Abouhalima, Mahmud, 182, 232
Abu-Musa, Ibrahim, 182, 232
Accurint, 280
Add123, 280
Airline Reporting Corporation, 218
Al Rashid Trust, 186, 233
Alpha MI, 92
Amazon.com, 103
America Online (AOL), 283
American Express, 215
American Institute on Domestic Violence, 131
Americans with Disabilities Act, 295
Ameridex, 281
Anderson, Chad, 128
Anti-Fraud Task Force, 225
Author Contact Information, 317
AutotrakXP, 281
Ayoub, Radwan, 182, 232
Bahour, Adnand, 232
Bank of America, 25
Bankers Trust, 212
Bardo, 273
Bay Industries, Inc., 233
Benavidez, Treyce, 84
Better Business Bureau, 288, 306
Bill of Rights, 308
Black Book Online, 283
Board of Elections, 277
Bodyguard, 4, 19, 255
bomb, 6, 23, 50, 76, 174, 283
Brinks, 268, 269
Brown, T. A., 67, 98, 105, 279, 283, 304, 314
Bureau of Export Administration, 184

Bureau of National Affairs, 130
Bürgel, 92
California Insurance Code, 190, 193
California Insurance Frauds Prevention Act, 190, 192
Calvert, James, 128
Carter, Dean, 128
Castleman, Scott, 132, 207
CDB Infoteck, 281
Census Reports, 295
Chamber of Commerce, 62, 92, 97, 288, 306
China Power Lighting Ltd., 297
Chinese Fishing Export Ltd., 297
ChoicePoint, 281
Christensen, Larry, 184
Civil Rights Act, 141
Clairborne, Liz, 128
Clinton, Bill, 274
Coalition Against Insurance Fraud, 224
Cohen, Joseph, 225
Collier, Brenda, Esq., 260
Confi-chek, 281
Conning & Company, 224
Conning Corporation, 224
Cook, Gary R., 20
Copernic, 297
Corinthia Group of Companies, 233
Creditreform, 92
Culligan, Joseph J., 268
Dallas Information Systems, 281
Damon Clinical Laboratories, 212
Data Protection Act, 91, 94, 97
Data-trac, 281
DBT, 89, 281, 282

DeKieffer, Don, 232
Department of Defense, 103
Department of Insurance, 192
Department of Justice, 287
Dow Jones, 288
Dun and Bradstreet, 92, 93
E Government, 95
Earnshaw, Joan, 256
EBAY, 198
eBlaster, 296
El Paso Information Center, 214
El-Al Airlines, 49
Elaws, 289
Electric Law Library, 290
Elliott, John M., 62
Equal Employment Opportunity Commission, 141
Equifax, 89, 93, 198
Ettisch-Enchelmaier, M., 91, 297
Eudora Pro, 123
Evidence Eliminator, 289
Experian, 198
Export Administration Bureau, 234
Exxon Corporation, 212
Fair Claims Practices Act, 193
Farley, Richard, 128
FBI, 66, 100, 185, 213, 216, 224, 233, 274, 277, 304, 316, 317
Federal Bureau of Prisons, 287
Federal Drug Free Workplace Act, 121
Federal Election Commission, 288
Federal Express, 265
FedEx, 95
Financial Crimes Enforcement Network, 214
findlaw, 315
First, Ruth, 8
Flat Rate, 279

Foreign Assets Control, 184, 234
Foreign Criminal Records, 289
Fortune 1000, 62
Fourth Amendment, 121, 122, 308, 309, 313
fraud, 6, 68, 86, 181, 182, 195, 198, 201, 207, 224, 232, 295
Freedom of Information Act, 99, 291
Gardner, Robert A., CPP, 110
Goldmann, Peter, 182
Google, 34, 293
Habash, George, 232
HackerWacker, 294
Hamas, 54, 233
Hanafy, Ibrahim Elsayed, 233
Hawkins, Mike, 308
Hill, Anita, 141
Homeland Security, 46, 98, 178, 294
Honda, 33
hostage, 161, 178
Hotmail, 31, 198
Icap SA, 92
Immigration and Naturalization Service, 98
Insurance Research Council, 224
Intellicorp, 282
International Air Transport Association, 214, 218
International Intelligence Network, 307
iQ411, 282
iqdata, 94, 196, 250
IRBSearch, 280
IRS, 217, 221, 257, 258, 284
Islamic Jihad, 54, 57
Keltner, Brian, 197
Klein, Barbara, 100
Knowx, 94, 195, 250, 281
KnowX, 286
Knox, Debra, 103
Kuebler, David W., Col., 252

Labor and Resources
   Subcommittee, 120
Laden, Osama bin, 101
Landstraust Hf, 92
Lennon, John, 6
Lexis-Nexis, 282
Linux, 31, 33
Mace, 71, 72, 134
Macintosh, 33
Martindale Hubbell, 193
Mason Jars, 254
Masterfiles, 281
McAffee, 33
McCourt, Michael G., 46
McMurray, Kenneth, 128
Merlin, 279
Microsoft Word, 34, 123, 293
Murrah Federal building, 23
National Association Of
   Investigative Specialists,
   307
National Council Of
   Investigation & Security
   Services, 307
National Retail Merchants'
   Association, 202
Nationwide Court Directory,
   295
Naylor, Rick, 128
Neighborhood Watch, 13
Neuss, 92
Nolo Press, 156
Northrop, 212
Norton Systemworks, 126
offshore banking, 294
One Liberty Plaza, 51, 58
Online 2005, 95
Orwell, George, 308
OSHA, 119
Outlook Express, 31, 38, 123
Pacer, 281
Palestinian Liberation Front,
   232
Palestinian Liberation
   Organization (PLO), 53
PallTech, 281
Pankau, Edmund J., 157,
   203, 213, 263

Party of God, 57
Pasteur, Louis, 82
PC Magazine, 285
Pentagon, 51, 182, 185
Pfizer Inc., 212
Press Association, 95
Privacy Foundation, 295
Protector Plus Voice Dialer
   Security System, 283
Pruitt, Alan, 103
Public Records Resources,
   282
PublicData, 280
Quality Shoes Company, 233
QuickInfo, 281
Radio Shack, 68
Rahn, Charles T., 106
ReferenceUSA, 294
Repa, Barbara Kate, 141
Research & Retrieval, 292
Research Data, 293
Riddle, Kelly E., 194, 246
Ripa, Kevin J., 30, 78, 88,
   123
Roberts, David P., 4, 49
Royal Caribbean Cruises Ltd.,
   212
Russell, Jeffrey A., 224
Savvy Data, 281
Schmedlen, Roger H., 200
Search Systems, 281, 285
Secret Service, 185, 232, 304
Sheffer, Todd, 232
Shiites, 57
Siciliano, Robert, 69
Simpson, James, 128
Slovo, Joe, 8
Smiley, Christina, 238
Source Resources, 282
South African Police Security
   Branch, 8
Spear, Robert K., 161
Specialty Technical
   Publishers, 48
Spence, Justin, 120
Swiss Telecom, 91
Systematic Alien Verification
   for

Entitlements/Employment, 98
Terrorist, 28, 52, 55, 56, 166, 182, 232
theft, 6, 24, 68, 85, 132, 140, 181, 186, 197, 200, 207, 296, 312
Thomas Publications, 140
Thomas, Judge Clarence, 141
threat, 4, 19, 32, 46, 52, 54, 59, 115, 117, 118, 131, 157, 161, 169, 219
Tigris Trading, Inc., 233
Trade Compass, 185, 234
Trans Union, 89, 198
Transactional Records Access Clearinghouse, 291
Treasury Enforcement Computer Service, 214
Trendsetter Barometer, 63
Uniform Commercial Code, 296
Union Carbide, 211
UPS, 95

US Find, 281
Walkman, 8
Wal-Mart, 154
WestLaw, 291
White House, 185
White, Charles Lee, 128
White-Collar Crime Fighter, 184, 186
Widell, Coleen, 128
Witherspoon, Ralph, 23
Wold, Ken, 299
World Association of Detectives, 97
World Bank, 25
World Convention, 105
World Trade Center, 23, 49, 51, 182, 185, 232
World Trade Organization, 25
XpertSearch, 281
Yahoo, 31, 198, 297
yellow pages, 304
Zalma, Barry, Esq., 187
Zone Alarm, 42

Give the Gift of BUSINESS SECURITY
To Your Supervising Employees, Colleagues, Clients, or Associates

YES, I want _____ copies of *Business Security* for $29.95 each plus $4.95 shipping and handling per book. Nevada residents please add $2.25 sales tax per book. International Orders are $39.95 each plus $9.95 shipping and handling per book.

My check or money order for $_____ is enclosed.
International orders must be accompanied by a postal money order in U.S. funds.

Please Print Plainly:

Name _____

Organization _____

Street Address _____

City/State/Zip _____

Country _____

Phone (____) _____ E-Mail _____

                Mail to:
                Crary Publications
                P.O. Box 42422
                Las Vegas, NV 89116-2422

Credit card ordering by phone: 1-877-734-7638. International callers should dial +1-770-319-2718. With all telephone orders you must include the product number **60730** with your request.

Satisfaction Guarantee and Return Policy

If you bought any book from us and paid 80-100% of the original list price before shipping, you may return the book in salable condition with your receipt within 90 days and we will issue a prompt, courteous refund (the shipping charge is non-refundable).

                www.BusinessSecurity.org
            E-Mail: order@businesssecurity.org

Save $6 off regular purchase price per book with this coupon on your next order of 2 or more copies of *Business Security*. Mail this coupon with your check or money order directly to the publisher listed above for your discount savings.